International Conference for Free and Open Source Software for Geospatial (FOSS4G 2014)

OSGeo Journal Volume 14, Issue 1

Portland, Oregon, USA
8-13 September 2014

Editors:

Barend Köbben

ISBN: 978-1-7138-1900-4

Printed from e-media with permission by:

Curran Associates, Inc.
57 Morehouse Lane
Red Hook, NY 12571

Some format issues inherent in the e-media version may also appear in this print version.

OSGeo
9450 SW Gemini Dr. #42523
Beaverton, Oregon
United States
97008

www.osgeo.org

Additional copies of this publication are available from:

Curran Associates, Inc.
57 Morehouse Lane
Red Hook, NY 12571 USA
Phone: 845-758-0400
Fax: 845-758-2633
Email: curran@proceedings.com
Web: www.proceedings.com

Free and Open Source Software for Geospatial (FOSS4G) Conference Proceedings

Volume 14 *Portland, Oregon, USA* Article 1

2014

Volume Introduction Letter

Barend Köbben
ITC–University of Twente

Recommended Citation
Köbben, Barend (2014) "Volume Introduction Letter," *Free and Open Source Software for Geospatial (FOSS4G) Conference Proceedings*: Vol. 14 , Article 1.
DOI: https://doi.org/10.7275/R56H4FMJ
Available at: https://scholarworks.umass.edu/foss4g/vol14/iss1/1

FOSS4G · PDX · 2014

OSGEO Journal Vol. 14 – July 2015

From the Editor...

by Barend Köbben

Welcome to this issue of the OSGeo Journal, comprising five research papers selected from the submissions to the Academic Track of FOSS4G 2014, which took place in Portland (Oregon, USA), from 8 to 13 September 2014.

FOSS4G, the global conference for Open Source Geospatial Software, is not an academic conference. The core audience has always been the people who make up the *open source communities*: The people that develop, create and craft the open source geospatial software. The actual applications are the glue which binds the community together; the aim of the FOSS4G community is to enable and enfranchise anyone to harness the power of geo-spatial software, regardless of their economic status. To acknowledge this, and to not create an isolated, exclusive, part of the conference, we scheduled presentations of the papers clustered with other, non-academic, papers based on subject matter. By this, we hope to have generated attention for academic input in the community and to cross-pollenate with industry, developers and users.

This year, the Academic Track submissions were a bit disappointing in number, but fortunately not in quality. At the conference itself the AT track chairs had fruitful discussions with the authors. This system of having an extra iteration based on the presentation and personal contact proved to be very useful. What we finally ended up with was a selection of ten papers, out of which the reviewing team considered three candidates fitting contributions for the Wiley Journal "Transactions in GIS". Another seven were offered publication in the OSGEO Journal and five of these have ended up in this issue.

Phillip Davis reports on the extensive work done to create a new innovative geospatial curriculum built around open source software, to increase both the quantity and quality of geospatial workers. Although funded and founded in the USA, the whole FOSS4G community benefits from this work. Specifically the curriculum that was developed, and is shared under a Creative Commons license, is a welcome source of teaching material for educators worldwide.

In their paper on *UrbanSim2*, Fletcher Foti and Paul Wadell describe an open source software platform for agent-based geospatial simulation, focusing on the spatial dynamics of urban development. This scientific tool library is an excellent case study for the power of combining open source work in the scientific programming community, to avoid having to build customized solutions in each domain.

Another example of the use of FOSS4G in scientific work is the paper of Jeffery Cavner *et al.* They have added phylogenetic capabilities (for the description of ecological processes, calculating and mapping biodiversity indices and such), to several open source platforms: As a QGIS plugin, and as web services, in the form of WPS algorithms.

Web Processing Services (WPS) are also the subject of the work of Ebrahim Poorazizi and Andrew Hunter, who report on an extensive analysis of five WPS servers (52° North, Deegree, GeoServer, PyWPS, and Zoo). They performed a quantitative analysis of the performance, as well as qualitative metrics such as software architecture, perceived ease of use, flexibility of deployment, and quality of documentation.

Finally, the paper on "GRASS GIS, Star Trek and old Video Tape" is a reflection of one of the highlight talks of the conference. The restored video of William Shatner explaining the virtues of GRASS version 2.0 in 1987 caused quite a stir, and in his paper Peter Löwe explains the importance of this video and the preservation process.

I'd like thank my fellow AT chair, Franz-Josef Behr, and the reviewers (listed on the imprint page at the end of this issue) for making the Academic Track and this journal issue possible.

Barend Köbben, ITC–University of Twente

FOSS4G2014 Proceedings Editor OSGeo Journal
http://wiki.osgeo.org/wiki/BarendKobben
b.j.kobben@utwente.nl

Contents

Educating 21st Century Geospatial Technology Industry Workers with Open Source Software

by Phillip Davis

Del Mar College (USA). pdavis@delmar.edu

Abstract

The global geospatial technology industry, in a study by UK-based Oxera commissioned by Google in January 2013, has been estimated at $150 USD billion to $270 USD billion per year ($110 billion euro to $199 billion euro). In a similar US-focused study, also commissioned by Google in 2013, the Boston Consulting Group (BCG) found the geospatial services industry employs approximately 500,000 people and generates around $75 (USD) billion in annual revenue ($55 billion euro). By any measure, the geospatial industry is large one, in both the US and globally. With this explosive growth, combined with the current generation of geospatial workers nearing retirement age in the next decade, it has become imperative to increase the number of well-qualified graduates from higher education programs, knowledgeable in the latest geospatial technology. This report describes one effort in the US to increase both the quantity and quality of these workers through the use of a new innovative geospatial curriculum built around open source software.

Keywords: Curriculum, open source, FOSS4G, education, university, college, QGIS, GTCM, DACUM, teaching, learning, ESRI, Google, FOSS.

1 Demand for Geospatial Industry Workers

The global geospatial technology industry, in a recent study by UK-based Oxera commissioned by Google in January 2013, has been estimated at $150 USD billion to $270 USD billion per year ($110 billion euro to $199 billion euro). In a similar US-focused study, also commissioned by Google in 2013, the Boston Consulting Group (BCG) found the geospatial services industry employs approximately 500,000 people and generates around $75 (USD) billion in annual revenue ($55 billion euro). By any measure, the geospatial industry is large one, both in the US and globally. With many of the current generation of the world's geospatial workers nearing retirement age in the next decade, it has become imperative to increase the number of well-qualified graduates from higher education programs, knowledgeable in the latest geospatial technology, to replace retiring workers and to meet the demand for even more workers in this expanding industry. Table 1 depicts this demand in the United States of America.

2 Limited Software Options for Students

In the US, software from a single vendor is used almost exclusively by 90% of the 1400 colleges and university academic GIS programs nationwide. By focusing heavily on the US higher education market for more than two decades, this vendor can legitimately claim their product is used to train 9 out of every 10 graduates in the US. Although their nonacademic business share of the global market was estimated at 40% in 2012, they dominate the US higher education section by a disproportionally large margin. This lopsided representation, we believe, is detrimental to the global competitiveness of US workers, as well as limiting their technical skill set. By providing a robust and well marketed GIS education program, this vendor dominates the academic GIS market. In the latest survey of US academic GIS departments nationwide (GeoTech Center; annual report, 2012), only 5% of colleges and universities reported offering any form of open source geospatial software. This same survey revealed that more than half of the faculty responding indicated an interest in using open source in their classrooms and labs.

3 What Is Lacking in Geospatial Software Instruction?

In order to provide students with the opportunity to work and gain competence in open source geospatial software, we must first build instructors an Open Source Software Learning Infrastructure. This in-

Occupation	Employment (2010)	Projected Growth (2010-2020)	Projected Growth Rate (2010-2020)
Geospatial Information Technician	210,000	51,600	3 to 9%
Remote Sensing Scientists and Technologists	30,000	13,300	3 to 9%
Remote Sensing Technicians	62,000	33,500	10 to 19%
Geodetic Surveyors*	51,000	24,200	20 to 28%
Mapping Technicians	57,000	20,000	10 to 19%
Cartographers and Photogrammetrists	14,000	6,100	20 to 28%
TOTALS	**424,000**	**148,700**	**(3 to 28%)**

Table 1. Geospatial Occupations U.S. Department of Labor Employment and Training Administration Projected Geospatial Job Grow (DOLETA). Source: U.S. Department of Labor Employment and Training Administration, O*NET Online, `http://online.onetcenter.org/`, September 1, 2013

frastructure must be as equally robust as the current leading commercial vendor's framework. This framework is able to offer a complete, ready-to-use curriculum and support products to academics at competitive pricing. This support includes: a) textbooks and lab manuals for all levels of learners, b) a virtual campus of online courses, c) professional development for educators through workshops, and d) robust community of practice through regional user's group and conferences. Compare this to the open source educational resources, where there is a lack of ready-to-use curriculum, limited opportunities for professional development, limited number of textbooks and online courses, and a small, but growing community of practice among educators using open source software. We cannot fully achieve the Geo-For-All initiative of the ICA-OSGeo pact without this infrastructure. The goal of our research is to increase the quantity and quality of open source software curriculum resources available to higher education faculty in order to boost its broader adoption in colleges and universities across the United States.

4 Determining the Worker's Knowledge, Skills and Abilities

The US Department of Labor's Geospatial Technology Competency Model (GTCM) is the recognized standard in defining the requisite skills of the industry workforce. This model provides a comprehensive list of the knowledge, skills, and abilities (KSA) required of workers in the geospatial technology industry. The model is represented as a pyramid, with the most fundamental skills at the base and building upward into more specialized knowledge areas. This model (Figure 1) has been used by hundreds of educators in the US, Europe and Asia to align GIS courses and curriculum with industry-identified KSAs. Beginning at the lowest tier (1), the foundational knowledge and skills are defined and applicable to all levels of workers in the industry—from entry-level technicians to expert scientist. Moving up in Tiers 2 and 3 more broad academic (Tier 2) and workplace (Tier 3) skills are defined, again applying to all workers in the industry. At Tier 4 we begin to define the foundation geospatial competencies required of all workers in the field. At Tier 5, the model separates into three broadly defined sectors of the industry, each with its own specific set of competencies germane to workers in that particular sector of the industry. The genius of the GTCM in Tier 5 was achieving, for the first time, a broadly accepted definition of the sectors. Moving into Tier 6, Occupational Specific Competencies, the model defines job-specific tasks and skills needed by those workers. These jobs are defined by the Department of Labor's Standard Occupation Codes, which are updated periodically, separate from the GTCM. Complimenting Tier 6 is the Geospatial Technology Management Competencies, which have been defined by the Department of Labor through the work of URISA in 2012. This Tier defines the broad management skills needed to organize and management significant geospatial projects and departments.

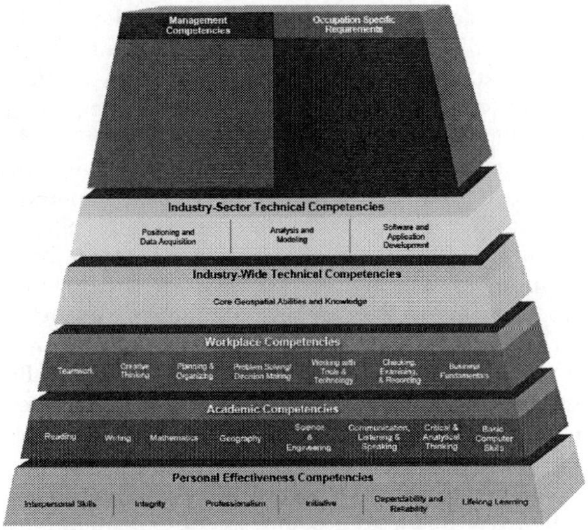

Figure 1. The US DOL Geospatial Technology Competency Model (GTCM). `http://www.careeronestop.org/competencymodel/pyramid.aspx?geo=Y`.

By contrast with the UCGIS Body of Knowledge (BoK), published in 2006, the Geospatial Technology Competency Model (GTCM) is the result of industry-driven input. The BoK, by comparison, contains some 1660 individual items and was created exclusively by a group of distinguished academics from Tier 1 research universities in the US. The GTCM contains a more finite 660 items, and represents the consensus outcome of a dozen industry-recognized experts in a two day panel that was conducted in March 2010. This panel of distinguished professionals represented the broadest possible cross-section of the industry, including surveyors, cartographers, geographers, computer scientists, remote sensing, photogrammetrist, and GNSS satellite experts. Their work was facilitated under the auspices of the US Department of Labor by a professional trained in the consensus-building process. The collective results of the GTCM national panel were then further vetted, during April 2010, among a much larger group of US Geospatial Industry professionals. This vetting included participation from national professional organizations, such as the American Association of Geographers (AAG), the American Society of Photogrammetry and Remote Sensing (ASPRS), and others. Through electronic surveys and public commenting, the GTCM was further refined in May 2010. Finally, the GTCM was reviewed and approved for publication by the US Department of Labor on June 10, 2010.

5 Localizing the Model with Regional Input

To make the national GTCM, containing 660 items, even more usable by educators, the GeoTech Center undertook a series of industry-led workshops around the country. These facilitated panels performed what is known as a DACUM, or Developing A Curriculum building exercise. Similar to the process used in the national GTCM panel, these regional DACUM panels, consisting of 6 to 12 professional, were limited to industry worker participation. The KSAs identified in these workshops were then vetted among a larger group of GIS professionals in the region using electronic surveys. These results were then finalized and published on the GeoTech Center website. While academics were allowed to observe, they were prohibited from participating in the actual workshop, assuring the results represented only industry-derived KSAs. The workers participating in the DACUM panels and electronic surveys included government workers, engineering technicians, GIS managers, etc. The results of these individual DACUM panels, held at five different locations between 2009 and 2012. These results were then collated and mathematically ranked using regression analysis to arrive at a final meta-analysisor MetaDACUM. The final report was peer-reviewed and published in the URISA Journal article 2010 no.2: p55-72 (`http://www.freepatentsonline.com/article/URISA-Journal/253845098.html`).

6 From Model to Material

The final step in making the GTCM and Meta-DACUM analysis relevant for educators was a two-year long curriculum-building effort that engaged more than 50 higher-educators from two year colleges and four year universities. In a collaborative effort, the 660 KSAs found in the GTCM and Meta-DACUM were further refine into a more definitive 330 individual KSAs, ranked according to importance and categorized into a model program of study (POS) and course level student learning outcomes (SLOs). Utilizing the proven methodology of facilitated group feedback and refinement used in the GTCM and DACUM workshops, the GeoTech Center, under direction of this author, convened a series of five educator workshops in 2011 and 2012 that produced the GTCM Model Certificate and Courses. These course outlines contain the basics of: a) syllabus, b) student learning outcomes (SLO), objective question assessments, and resource recommendations, provides a basis for the design of new GIS curriculum that reflects the true state-of-the-art in current geospatial technology, as defined by industry, and interpreted by academics. This GTCM Model Certificate and Course recommendation is published at `http://www.geotechcenter.org/gtcm-curriculum-guide-20.html`.

7 Final Step in Curriculum Development

These Model course outlines became the basis for further curriculum development work directed by the author for the US Department of Labor National Information, Security & Geospatial Technology Consortium (NISGTC) between June 2012 and June 2014. The result is a complete set of GIS courses, including lecture and laboratory curriculum materials, for five complete GIS courses. These courses include: a) GST 101—Introduction to GIS, b) GST 102—Spatial Analysis, c) GST 103—Data Acquisition and Management, d) GST 104—Cartography, and e) GST 105—Remote Sensing, represent the model program of study for a GIS Technician and are shown in Table 3.

Course Number	Course Title
GST 101	Introduction to Geospatial Technology
GST 102	Spatial Analysis
GST 103	Data Acquisition & Management
GST 104	Cartographic Design
GST 105	Remote Sensing

Table 3. National Model Certificate GTCM FOSS4G Courses.

These courses are now complete and ready for distribution under the Creative Commons 3.0 license allowing for the free use and redistribution with attribution (`http://nter.riosalado.edu`). While the lecture portion of these five courses is generic and applicable to any software implementation, the initial laboratory portion was built using Esri ArcGIS 10.1 proprietary software, as required by our contract award with the US federal funding agency, the US Department of Labor.

In 2014, we have completed a series of complimentary set of GIS laboratory experiences based around the latest Quantum GIS (QGIS 2.2) software build. With this complete set of courses, labs, and support material, we are now prepared to begin building the open source geospatial educational infrastructure to develop a global community of practice among GIS educators worldwide. Beginning with their debut at the International FOSS4G 2014 Conference in September 2014, we will be prepared to launch a national initiative to increase the adoption of open source geospatial software in colleges and universities across the US. It is our goal to both compliment the proprietary software in existing GIS programs, as well as assist those colleges and universities desiring to start new GIS academic programs based on the open source software model. By leveraging the rapidly expanding ICA-OSGeo Open Source Software Laboratory Network, we will be offering our curriculum free of charge under the Creative Commons BY 3.0 license. This Geo-For-All initiative will further our commitment to bringing the latest possible technology learning experience to our students on a global scale.

References

A Massive Open Online Course. The Georgia Tech MOOC. Blog at WordPress.com, n.d. Web. 17 Jul. 2012. http://gtmooc.com/

An Open Badge System Framework: A foundational piece on assessment and badges for open, informal and social learning environments. DML Central. n.p., 1 Mar. 2012. Web. 22 May 2012. http://dmlcentral.net/resources/4440

Bates, Tony. The challenge of converting MOOCs (or anything else) into college credits. Online Learning and Distance Education Resources. Contact North, 8 Aug. 2012. 10 Aug. 2012. http://www.tonybates.ca/2012/08/08/the-challenge-of-converting-moocs-or-anything-else-into-college-credits/

Beum, S. G. & Im, J. H. (2012). A Case study of UNIST e-education: Inverted (Flipped) learning model. 2012 Spring Conference of the Society of e-Learning, 3(1), 190-194.

Carnevale, Anthony, Stephen Rose, and Andrew Hanson. Georgetown University. Center on Education and the Workforce, 5 Jun. 2012. Web. 14 Jun. 2012. http://cew.georgetown.edu/certificates/

Creative Commons. Center for the Public Domain. n.p., n.d. Web. 9 May 2012. http://creativecommons.org/

Davis, Josh. Mil-OSS. Open Source for America. n.p., n.d. Web. 1 May 2012. http://www.mil-oss.org/

Dempsey, Caitlin. Where are the GIS Jobs? A Look at the GIS Job Market in the United States. GIS Lounge. GIS Lounge, 19 Jun. 2012. Web. 28

Jun. 2012. http://gislounge.com/where-are-the-gis-jobs-a-look-at-the-gis-job-market-in-the-united-states/

Donert, Karl. Does anyone have any thoughts on the role that Centres of Excellence play in knowledge transfer and transmitting innovative education approaches? Linked In. Linked In Corporation. Web. 14 May 2012

Fain, Paul. College Credit Without College. Inside Higher Ed. n.p., 7 May 2012. Web. 14 May 2012. http://www.insidehighered.com/news/2012/05/07/prior-learning-assessment-catches-quietly

Fain, Paul. Making it Count. Inside Higher Ed. n.p., 15 Jun. 2012. Web. 6 Aug. 2012. http://www.insidehighered.com/news/2012/06/15/earning-college-credit-moocs-through-prior-learning-assessment

Ferenstein, Gregory. Move Over Harvard And MIT, Stanford Has The Real Revolution In Education. Mid-Pacific ICT Center. City College of San Francisco. Web. 2 Aug. 2012. http://mpictcenter.blogspot.com/2012/06/move-over-harvard-and-mit-stanford-has.html

FOSS4G Workshop for Educators at FOSS4G. Bird's Eye View GIS. Bird's Eye View GIS, 13 Jul. 2011. Web. 1 May 2012. http://www.birdseyeviewgis.com/blog/2011/7/13/foss4g-workshop-for-educators-at-foss4g.html

Friedman, Thomas L. Come the Revolution. The New York Times. The New York Times Company, 15 May 2012. Web. 18 May 2012. http://www.nytimes.com/2012/05/16/opinion/friedman-come-the-revolution.html

Gadja, R. (2004). Utilizing Collaboration Theory to Evaluate Strategic Alliances. American Journal of Evaluation 25(1): 65-7.

Gakstatter, Eric. Open Source GIS: Part II. Geospatial Solutions. North Coast Media, LLC., 24 May 2012. Web. 19 Jul. 2012.

Gakstatter, Eric. Reader Response on Open Source and Mobile Devices. Geospatial Solutions. North Coast Media, LLC., 12 Jul. 2012. Web. 23 Jul. 2012.

Hill, Phil. The Emerging Landscape of Educational Delivery Models. eLiterate. WordPress, 15 Mar. 2012. Web. 10 Apr. 2012. http://mfeldstein.com/the-emerging-landscape-of-educational-delivery-models

Hill, Phil. The Master Course: A Key Difference in Educational Delivery Methods. eLiterate. WordPress, 22 Mar. 2012. Web. 18 Apr. 2012. http://mfeldstein.com/the-master-course-a-key-difference-in-educational-delivery-methods/

Ivy League education free on the web. Click: The World of Technology Across the BBC. BBC News. 5 May 2012. http://news.bbc.co.uk/2/hi/programmes/click_online/9720231.stm

Reiser, John. FOSS4G-NA software for education BOF. EVERNOTE. n.p., 11 Apr. 2012. Web. 24 May 2012. http://chronicle.com/article/How-One-Instructor-Teaches/131656/?sid=pm&utm_source=pm&utm_medium=en

Kamenetz, Anya. How Coursera, A Free Online Education Service, Will School Us All. Fast Company. Mansueto Ventures LLC, 8 Aug. 2012. Web. 10 Aug. 2012. http://www.fastcompany.com/3000042/how-coursera-free-online-education-service-will-school-us-all

Kim, Joshua. Playing the Role of MOOC Skeptic: 7 Concerns. Inside Higher Ed. n.p., 21 May 2012. Web. 14 Jun. 2012. http://www.insidehighered.com/blogs/technology-and-learning/playing-role-mooc-skeptic-7-concerns

Kolowich, Steve. MOOCs and Machines. Inside Higher Ed. n.p., 10 May 2012. Web. 14 May 2012. http://www.insidehighered.com/news/2012/05/10/candace-thille-talks-moocs-and-machine-learning

Kolowich, Steve. Proto-MOOC Stays the Course. Inside Higher Ed. n.p., 24 Apr. 2012. Web. 14 May 2012. http://www.insidehighered.com/news/2012/04/24/opencoursedigital-storytelling-enjoys-modest-success

Kolowich, Steve. The Online Pecking Order. Inside Higher Ed. n.p., 2 Aug. 2012. Web. 6 Aug. 2012. http://www.insidehighered.com/news/2012/08/02/conventional-online-universities-consider-strategic-response-moocs

Kouyoumjian, Victoria. Open Source, Closed Source: Moving to the Middle. Esri. n.p., 24 Oct. 2011. Web. 13 Apr. 2012. http://blogs.esri.com/esri/esri-insider/2011/10/24/open-source-closed-source-moving-to-the-middle/

Large Turn-Out Demonstrates Government's Interest in Open Source Software. Geospatial Solutions. North Coast Media, LLC., 13 Jun. 2012. Web. 29 Jun. 2012.

Love, Julia. U.S. Opens a Door to a Dream for Young Illegal Immigrants. The Chronicle of Higher Education. n.p., 15 Aug. 2012. Web. 15 Aug. 2012. http://chronicle.com/article/US-Opens-a-Door-to-a-Dream/133649/

Mangan, Katherine. In This Online University, Students Do the Teaching as Well as the Learning. The Chronicle of Higher Education. n.p., 18 Jun. 2012. Web. 21 Jun. 2012. http://chronicle.com/article/In-This-Online-University/132307/

McCormick, Coleman. Conversations about open source and higher education at FOSS4G. Spatial Networks. Spatial Networks, Inc. – Location Leverage, 2011. Web. 11 Jul. 2012. http://spatialnetworks.com/blog/2012/04/conversations-about-open-source-and-higher-education-at-foss4g/

McKendrick, Joe. Cloud CoUDL Cut $12 Billion from US Government Annual Deficit: Study. Forbes. Forbes.com LLC, 30 Apr. 2012. Web. 5 Jun. 2012.

Menke, Kurt. Experiences Teaching Free and Open Source GIS at the Community College Level. Directions Magazine. Directions Media, 7 Nov. 2011. Web. 3 May 2012. http://www.directionsmag.com/articles/experiences-teaching-free-and-open-source-gis-at-the-community-college/212522

Meyers, Rhiannon. Coastal Bend colleges, universities recognized for graduating Hispanics in high-demand fields. Caller.com. The E.W. Scripps Co., 2 Aug. 2012. Web. 10 Aug. 2012. http://www.caller.com/news/2012/aug/02/area-colleges-and-universities-recognized-for-in/

Open Source Computer Lab. UMassAmherst. University of Massachusetts Amherst, 2007. Web. 11 Jul. 2012. http://www.umass.edu/opensource/

OSBC 2011 Survey Results. The Future of Open Source. Drupal Gardens, n.d. Web. 10 Jul 2012. http://www.futureopensource.net/osbc-2011-survey-results

Parry, Marc. Supersizing the College Classroom: How One Instructor Teaches 2,670 Students. The Chronicle of Higher Education. n.p., 29 Apr. 2012. Web. 8 Aug. 2012. http://chronicle.com/article/How-One-InstructorTeaches/131656

Quinn, Clark. MOOC Reflections. Learnlets. WordPress. 29 Feb. 2012. Web. Date of Access. http://blog.learnlets.com/?p=2562

Segal, Tom. Rethinking the Learning Experience: Part IV. The Huffington Post. The Huffington Post.com Inc., 9 Aug. 2012. Web. 10 Aug. 2012. http://www.huffingtonpost.com/tom-segal/post_3750_b_1755869.html

Stansbury, Meris. Universal Design for Learning: The next big thing in school reform? eSchool News. eSchoolMedia and School News, 16 May 2012. Web. 18 May 2012. http://www.eschoolnews.com/2012/05/16/universal-design-for-learning-the-next-big-thing-in-school-reform/

Stojic, Mladen. Are Defense Cuts Good for the Geospatial Sector? Directions Magazine. Directions Media, 26 Apr. 2012. Web. 3 May 2012. http://www.directionsmag.com/articles/are-defense-cuts-good-for-the-geospatial-sector/247070

Stommel, Jesse. The March of the MOOCs: Monstrous Open Online Courses. HYBRID PEDAGOGY: A Digital Journal of Teaching & Technology. Hybrid Pedagogy, 23 Jul. 2012. Web. 31 Jul. 2012. http://www.hybridpedagogy.com/Journal/files/MOOC.html

Trumbull, Mark. Harvard and MIT to offer online courses. A step in lowering college costs? The Christian Science Monitor. The Christian Science Monitor, 2 May 2012. Web. 24 May 2012. http://www.csmonitor.com/USA/Education/2012/0502/Harvard-and-MIT-to-offer-online-courses.-A-step-in-lowering-college-costs

Watkins, S. Craig. From Theory to Design: Exploring the Power & Potential of 'Connected Learning', Part One. DML Central. n.p., 2 Aug. 2012. Web. 10 Aug. 2012.

Watters, Audrey. 5 Things I've Learned From MOOCs About How I Learn. Inside Higher Ed. n.p., 9 May 2012. Web. 14 May 2012. http://www.insidehighered.com/blogs/hack-higher-education/5-things-ive-learned-moocs-about-how-i-learn

Watters, Audrey. Udacity's CS101: A (Partial) Course Evaluation. Inside Higher Ed. n.p., 14 Apr. 2012. Web. 14 May 2012. http://www.insidehighered.com/blogs/hack-higher-education/udacitys-cs101-partial-course-evaluation

What is a MOOC? Report. 8 Dec. 2010. YouTube. 6 Aug. 2012. http://www.youtube.com/watch?v=eW3gMGqcZQc

White, Irenie. Open Source Drives Software Innovation. Credativ. n.p. 11 Apr. 2012. Web. 1 May 2012. http://blog.credativ.com/en/2012/04/open-source-drives-software-innovation.html

Young, Elise. Social Network for Class of 400,000. Inside Higher Ed. n.p., 19 Jun. 2012. Web. 9 Jul. 2012. http://www.insidehighered.com/news/2012/06/19/stanford-adds-social-learning-component-free-online-course

Young, Jeffrey. A Conversation With Bill Gates About the Future of Higher Education. The Chronicle of Higher Education. n.p., 25 Jun. 2012. Web. 2 Aug. 2012. http://chronicle.com/article/A-Conversation-With-Bill-Gates/132591/

Young, Jeffrey. 4 Professors Discuss Teaching Free Online Courses for Thousands of Students. The Chronicle of Higher Education. n.p., 11 Jun. 2012. Web. 2 Aug. 2012. http://chronicle.com/article/4-Professors-DiscussTeaching/132125

MOOCs' Little Brother September 6, 2012. By Steve Kolowich Read more: http://www.insidehighered.com/news/2012/09/06/u-maine-campus-experiments-small-scale-high-touch-open-courses

Inside Higher Ed. edX announces option of proctored exam testing through collaboration with Pearson VUE https://www.edx.org/press/edX-announces-proctored-exam-testing

Wikle, T. 2011. Planning Considerations for Online Certificates and Degrees in GIS, URISA Journal 22(1): 21-30

Rose, Universal Design for Learning. ERIC Resources Clearinghouse.

UrbanSim2: Simulating the Connected Metropolis

by Fletcher Foti and Paul Waddell

University of California, Berkeley (USA). fscot-tfoti@gmail.com

Abstract

UrbanSim is an open source software platform for agent-based geospatial simulation, focusing on the spatial dynamics of urban development. Since its creation UrbanSim has been used in the official planning processes for at least a dozen regional governments which were used to help allocate billions of dollars in regional investments in transportation infrastructure.

UrbanSim was first conceptualized in the late 1990's and implemented using the Java programming language. The technology landscape for scientific computing changed dramatically after that, and by 2005 UrbanSim was converted to Python, making heavy use of Numpy to vectorize calculations. By 2014, it became clear that UrbanSim should be reimplemented again to take advantage of significant advances in the libraries available for scientific Python. The new version of UrbanSim, called UrbanSim2, makes extensive use of community-supported scientific Python libraries to reduce the amount of domain-specific customized code to a minimum.

UrbanSim is an excellent case study for the power of leveraging the work of the scientific programming community as scaffolding for a domain-specific application, as opposed to building an extensive customized solution in each domain. Additionally, the open and participatory nature inherent in nearly all of the open source projects described here has been particularly embraced by governments, who are often reticent to support large commercial institutions and balkanized and private data formats and software tools.

Keywords: Open Source; Regional Planning; City Planning; Transportation Planning;.

1 Introduction

UrbanSim is an open source software platform for agent-based geospatial simulation, focusing on the spatial dynamics of urban development. It simulates the choices of locations of households and businesses in a metropolitan region in order to predict demand for public infrastructure such as transportation, energy and water. It has been most widely used for regional transportation planning, to assess the impacts of transit and roadway projects on patterns of urban development, and the indirect effects these have on travel demand. In recent years, urban models are increasingly used to understand how to reach sustainability goals, including reducing resource use and land consumption, and how best to substitute sustainable modes like transit, walking, and biking for increasing automobile use.

UrbanSim was first conceptualized and implemented almost 15 years ago (Waddell, 2000, 2002). The initial implementation was in Java, and for performance reasons it used an approach of 'exploded objects' to represent the millions of agents in its simulation, to minimize object overhead (Noth et al., 2003). By 2005, a decision was made to re-implement the UrbanSim platform in Python, taking advantage of the rapid advances made in the scientific Python community, most notably Numpy for multi-dimensional array computations. This version was referred to as the Open Platform for Urban Simulation (OPUS), and intended to stimulate broad collaboration as an open source project among international research teams working on urban modeling (Waddell et al., 2005).

Since its creation and release on the web as an open source project in 1998, UrbanSim has been increasingly used in the official planning processes for at least a dozen Metropolitan Planning Organizations (MPOs), which use UrbanSim and their travel model platforms to evaluate the planning of billions of dollars in regional investments in transportation infrastructure. By the end of its first ten years, UrbanSim had become the most widely used land use modeling platform for regional planning by MPOs (Lee, 2009), although a variety of other models have been used by MPOs, including 'home-grown' models (Hunt and Abraham, 2005; Wegener, 2004).

UrbanSim has grown over the past two decades to become a robust open source software platform, and its history closely parallels the history of many other open source projects, including the Python libraries on which it has grown to depend. Python was introduced by Guido van Rossum in 1991 and began to rapidly gain popularity. A key Python library -the vector and matrix manipulation library NumPy -was first released in 2006. When UrbanSim was reimplemented in Python beginning in 2005, a great deal of code was created using Numpy to build statistical models for estimating the parameters of regression and discrete choice models, and for efficiently simulating using the fitted models. In order to manage data effectively, the research team developed a heavyweight DataSet Class, to make sets of Numpy arrays behave more like relational database tables, with relational joins and a variety of query operations. In order to make the code more accessible to modelers, a domain-specific expression language was created (Borning et al., 2008). And in order to create a graphical user interface, an extensive PyQt infrastructure was built to auto-generate GUI elements from an XML file manipulated by the GUI. In short, a large volume of domain-specific, customized code accumulated as the project met a variety of needs. It eventually grew to over 150K lines of code, and debugging became more complex due to the many levels of indirection in the Python tracebacks whenever an error occurred. In other words, UrbanSim had become a large domain-specific application with a large customized code base to solve several of the problems that were eventually taken up by the scientific programming community as a whole. As a result, it was becoming a challenge to maintain.

In 2010, the Pandas data analysis library was released. Its user community has grown rapidly, and it has become a standard part of many scientific Python bundles. Pandas has implemented in a more general way most of the functionality that the UrbanSim team had developed in the UrbanSim DataSet class and the OPUS expression language. The new version of UrbanSim, called UrbanSim2, is the result of a reassessment of the landscape of available scientific Python libraries, and completely eliminates most of its legacy customized data management and expression language code, and its customized graphical interface. It replaced the UrbanSim DataSet class and expression language with Pandas, and replaced its GUI with Python, at least initially, with ongoing experimentation in web-based interfaces. The implementation of UrbanSim has thus again gone through a massive transformation in its software implementation, making an excellent case study for the power of leveraging the work of the scientific programming community to provide the scaffolding for a domain-specific application, rather than building and maintaining a heavy-weight customized solution in each domain.

This paper will outline this process and the argument for refactoring large domain-specific applications to rely more extensively on well-supported open source libraries such as exist in the scientific Python community. We begin by giving a brief history of urban modeling, followed by a discussion of the theory and implementation needs of UrbanSim, a description of the implementation efficiencies gained between the two versions of UrbanSim made by leveraging current open source geospatial projects, and closing with a discussion of future work on the topic.

2 A Brief History of Urban Modeling

Urban modeling began as economic theory, as early as 1826 in Von Thünen's "The Isolated State" (translated in Von Thünen and Hall, 1966), in which he outlined how in a "featureless plain" different crops would be grown in concentric rings around the city depending on their market value and cost of transportation. High value crops like vegetables and also heavy crops like firewood would both be grown near to the town while grain and ranch animals could be grown far from the city.

Walter Christaller then expanded these ideas to the urban context in Central Place Theory (Christaller, 1968) by proposing that while some products known as comparison goods —e.g. automobiles or appliances— would be consumed only rarely, other products known as convenience goods —e.g. food— would be consumed repeatedly. He proposed a geometry within the city of nested and overlapping hexagons in which vertices are shopping nodes and convenience goods would occur roughly 6 times as frequently within the city as comparison goods.

The modern era of urban models probably began with the exploration of the idea that increased transportation access leads to increased development intensity (a centralizing force), and that people tradeoff increased transportation costs to consume more housing (a de-centralizing force) (Hansen, 1959; Alonso, 1964). This presaged the field of transportation-land use interaction in which researchers explore how the built environment affects how people travel. Seminal work in the field shows that density, diversity, and design of the built environment of our cities all impact how far people will travel and which modes they will take, with more dense and diverse environments encouraging sustainable modes like walking and transit (Kockelman and Cervero, 1997; Ewing and Cervero, 2001).

Almost as soon as people understood that the built environment has a strong effect on how people travel, academics and transportation planners began to model the transportation systems of entire urban regions in order to predict travel patterns under counterfactual situations, like building a new highway or transit line, or accommodating an increase in population. The most prevalent methodology still in use today is the 4-step model (de Dios Ortúzar et al., 2001), in which the four steps are: trip generation, trip distribution, mode choice, and route choice. This framework runs four statistical models in sequence, answering for each person how many trips to take, where to take them, on which mode (auto, transit, walk, etc) and by which route.

Eventually planners began to ask more complicated questions of the models, including the impact of toll lanes and bridges, the effect of changing transit service characteristics, the result of carpooling and household coordination, and many other highly detailed policy questions. This gave birth to the modern advanced transportation models which fall under the rubric of activity-based travel models (ABMs) first proposed by Ben-Akiva and Bowman (Ben-Akiva et al., 1998; Bowman and Ben-Akiva, 2001).

In this paradigm, every person in a region is modeled as they move through the simulated day, sometimes for time increments as small as 5 minutes, capturing where each person is, what they are doing, and how they move from place to place. The dominant methodology in this framework is to run discrete choice models in sequence; the Portland implementation of the Ben-Akiva and Bowman framework has five levels of hierarchical choices: activity-patterns, time-of-day, mode-destination, sub-tours and intermediate stops.

Although the methodology employed by UrbanSim will be discussed in detail in the next section, the theory proposed therein owes direct lineage to this history of transportation models in the literature. UrbanSim also simulates a number of statistical models in sequence, using many of the same methods as the ABMs. In point of fact, many of the most advanced cities run both ABMs and land use models, using land use models to predict the spatial distribution of households and jobs (and other economic, environmental and social indicators) and the ABMs to predict the demand for transportation infrastructure and other travel characteristics.

Most urban modeling is performed at the level of regional government (i.e. MPOs), although large cities sometimes also have implementations of the models described here. MPOs are regional governmental bodies that were formed by the 1962 Federal-Aid Highway Act whose main purpose is to create and implement a long-term (typically 30 year) vision for the transportation infrastructure of a region, balancing the needs of constituent cities, as well as environmental and social equity considerations, while meeting demand for new highways and transit lines and maintaing existing infrastructure.

In California, state law SB-375 provides ground breaking legislation that due to the interconnected nature of land use and transportation (Barbour and Deakin, 2012), requires all MPOs to coordinate planning of land use and transportation infrastructure and to implement both land use models and transportation models. California MPOs are thus home to some of the most advanced urban models that exist today.

3 The Theory of UrbanSim

UrbanSim is built from several individual models of urban behavior. The models are typically statistically estimated, but this is not an essential requirement. In fact, UrbanSim can be viewed as a batch data analysis process with separate modules, each representing a specific urban behavior, and each module is allowed to read and write to the set of available urban data which includes at minimum: parcels for spatially subdividing land, the buildings which exists on those parcels, and the households and jobs which occupy that built space. The simply defined purpose of UrbanSim is to predict the spatial distribution of households and jobs in a future year, with an accurate representation of where new buildings will be built.

UrbanSim at its core is four models -the price model, location choice model, transition model, and real estate development model. Residential price models are called hedonics (Rosen, 1974; Waddell et al., 1993) and are usually linear regressions where the dependent variable is price (or a transformation of price) and independent variables typically include square footage, lot size, number of bedrooms and bathrooms, and attributes of the neighborhood like average income, regional accessibility by automobile and transit, local accessibility by walking, and others. A residential location choice model (McFadden, 1978; Lee et al., 2010) is a logit models where the number of alternatives is discrete and often quite large. For instance, households might choose from among all of the neighborhoods in a region in their choice of residence. Transition models are used to change the demographics of the population, including aging, family formation and separation, as well as births, deaths, and migration from other regions. As most regions in the United States are growing, the challenge becomes housing all the new households, and thus the creation of new residential buildings must also be modeled accurately.

The real estate development model is an extremely specialized model in Urban-Sim and is used to capture real estate developer behavior and analyze the cash flow of potential developments for profitability. This is called a pro forma (Miles et al., 2000; Brueggeman and Fisher, 1997) and is traditionally performed in a spreadsheet program, but UrbanSim uses Python to perform millions of pro formas for a large region to analyze the profitability of a multitude of possible buildings. Inputs to pro formas are rents or prices by unit type (1 bedroom, 2 bedroom, etc), construction costs per square foot, prevailing interest rates (and forecasts for future interest rates), the rate at which households will occupy a new development (called absorption), and jurisdictional policies including zoning restrictions, affordable housing policies, and parking requirements.

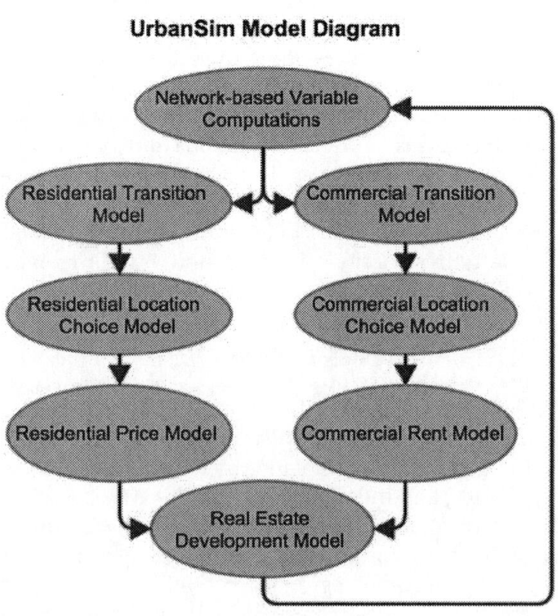

Figure 1: Diagram of the simplified UrbanSim model system.

Although the models above are described in detail for the residential/housing model set, there are analogous models for commercial entities including rent models for retail, office, and industrial building types, location choices for jobs by employment sector, predictions of job growth and decline, and the creation of new commercial buildings. The first three models are generally independent for the residential and commercial model sets (although aggregations of access to commercial uses might be variables in the residential models and vice versa). The real estate development model requires coordination between the residential and commercial markets; any parcel which is zoned to allow several potential uses must make a choice among alternatives based on the relative profitability of producing a building of a given type. Thus different uses compete for land with the highest profitability uses outcompeting less profitable uses. A diagram of the UrbanSim system of models is shown in Figure 1.

4 Network-based Spatial Variables

The selection of geography is enormously important in understanding any urban behavior. For instance, variables that are predictive of home prices and residential location choices can include the boundaries of school districts, other public goods provided within the boundary of a city (e.g. police protection), and these are large geometric shapes that are well-defined in the region. On the other hand, variables used to describe a person's perception of his neighborhood are not as easily defined, and much research has been performed to understand how people interpret their surrounding areas (Guo and Bhat, 2007; Grannis, 1998).

Although it is standard practice to use large polygons like census tracts, city boundaries, zip codes, etc to provide mutually exclusive boundaries for aggregations in the city, nonetheless this approach has clear weaknesses as polygon definitions are subject to judgement, they exhibit boundary effects (e.g. an element is either fully included or not included in an aggregation), and almost no spatial process will be completely homogeneous within such a polygon boundary. This can lead to the well documented MAUP (Modifiable Areal Unit Problem) (Openshaw, 1984), which is a potential issue for almost all of the spatial models used by UrbanSim.

To avoid this problem, UrbanSim now uses a framework for quantifying urban space where the city is represented as land use spatially located within a multi-modal transportation network (Foti, 2014). In this formulation, all land use is allocated to the nearest street intersection of the local street network and then aggregations can be performed within buffers surrounding every origin street intersection in the network. Figure 2 shows how parcels of land are each mapped to their nearest street node. In this figure, points represent intersections, lines represent the local streets obtained through OpenStreetMap, and parcels are assigned the color of the nearest street intersection.

Figure 2: Map of parcels to street nodes (mapping indicated by color) .

Aggregations are performed to a user-specified horizon distance along the network with a user-specified decay so that items that are further away are weighted so that they affect the origin location less, consistent with Tobler's second law of geography (Tobler, 1970). Aggregations can be any of the standard statistical measures including max, min, mean, standard deviation, and sum. Networks are also fully abstracted so that where information is available, local street networks, transit schedules, or congested (i.e. accounting for traffic) automobile travel times can be used to perform aggregations (see Foti, 2014 for more details on how these computations are performed). Thus variables can be computed such as "average income within 500 meters" or "jobs in the technology sector within 45 minutes transit ride," and many others.

The current implementation leverages the high performance network algorithm Contraction Hierarchies (Geisberger et al., 2008) and is written in C and fully multithreaded so that aggregations are computed very quickly. Local-scale aggregations (akin to WalkScore) can be performed for every street intersection in the United States in about 12 seconds. For a large region like the Bay Area, aggregations can be performed for all of the 226 thousand street nodes in a small fraction of a second.

The advantage of this "street node geography" is that its geometric definition is an emergent property of the local street network, which has a very real physical manifestation (i.e. is not subjectively defined). Additionally, the distance between intersections is usually small in dense urban areas which need to be represented most accurately, thus the land mapped to a street node is far more likely to be homogenous than the land within a geography as large as a census tract. Finally, network aggregations are overlapping and a decay is applied so that there are no boundary effects. Although representing actual access and egress points for parcels on the street network is possible, this information is not easily obtainable at this time, and street node geography provides a reasonable compromise in accuracy and also provides an order of magnitude increase in performance.

The current implementation does have a few limitations, including the lack of information about local-scale pedestrian access, sidewalks, and street crossings, information on the qualitative aspects of the pedestrian environment, etc, but is nontheless an extremely efficient substitute for polygon-based aggregations in GIS. The use of these variables is now ubiquitous in new UrbanSim implementations, and typically all network aggregation variables needed to run models are computed at the beginning of each simulated year so that all subsequent models will have access to these variables.

Although these metrics are extremely useful as independent variables in urban statistical modeling, it is hoped that this framework will be generally useful as a method for visualizing spatial data with far more precision than is typical with large polygonal geographies. For example, Figures 3 and 4 show the average home sales price in the Bay Area using zonal aggregations (Figure 3) and aggregations along the local street network (Figure 4). Although the color scale is discrete, the actual values for the aggregation are continuous and smooth for the network aggregations, and discontinuities are easily visible with the polygonal boundaries.

5 Leveraging Open Source Tools

UrbanSim was first translated to the Python programming language in 2006. In the intervening decade, the available technology and best practices for a large software project have changed dramatically. Innovations of particular relevance to the new UrbanSim implementation include:

- Python has added numerous supporting li-

Figure 3: Residential rent in the Bay Area aggregated by zone.

Figure 4: Residential rent in the Bay Area aggregated by network buffer queries.

braries including the Pandas data analysis package, Patsy, and SciKits and StatsModels for statistical analysis

- XML has been replaced by JSON for specifying key-value configuration documents
- The web has become the ubiquitous technology for graphical interfaces with advances including Angular, Leaflet, D3, Bootstrap, and others
- Github has become a free (for public projects) online collaborative tool ideal for large distributed code projects of this sort
- Anaconda now provides a well-tested Python distributions with numerous included Python packages to reduce installation headaches
- Spatial databases like PostgreSQL/PostGIS and the OSGeo packages have made GIS functionality available without dependency on commercial software packages like ArcGIS by ESRI

As the technology landscape has changed so dramatically, it became clear that UrbanSim would need to be overhauled substantially. In particular, the original UrbanSim was written in the year 2006, and the same needs that drove the eventual creation of the Pandas data analysis library were also present in the UrbanSim project. An abstraction layer was required to overlay a NumPy columnar datastore with names and types, the ability to read and write multiple formats of data, to merge/join different datasets, and to perform various aggregations across categorical variables to compute sums, means, and other typical statistical metrics. These operations became the basis for the UrbanSim expression language (Borning et al., 2008) and the similarities to the eventual Pandas package (McKinney, 2012) are many. Pandas has now been widely embraced in the Python community, with over 200 contributors, 9,000 commits, and at the time of this publishing 88,000 downloads per month.

The Pandas project is but one example of the integration with the larger open source software community that needed to take place, and so it was decided that UrbanSim would be re-implemented from scratch to work directly with Pandas, StatsModels, SciKits, etc. Although many of the more nuanced behavioral models have not yet been ported to the new framework, the bulk of the work has now been completed, and the new models have been used in active planning processes in the regions of Denver and Paris, and implementations are currently underway in many other regions. It should be noted that the use of well-tested community-supported frameworks has reduced the code complexity from over 100,000 lines of code to only 4,039 lines at the time of this writing. The new version of UrbanSim is now supported by the company Synthicity and is available as open source software under the AGPL license at https://github.com/synthicity/urbansim.

6 Statistical Model Configuration using JSON

Viewed as described in the previous sections, UrbanSim is essentially a configuration system for Pandas variables and statistical models provided by StatsModels. In fact, a small handful of parameters can describe each model configuration. In this section, the residential price hedonic will be used as the canonical example, and below is the list of parameters necessary to specify such a model:

- A small code wrapper is required to describe where the data comes from. Currently, tabular data is stored in the HDFStore as is common with Pandas.
- The main table for estimation/simulation must be specified. The canonical example would be a data table of home prices with sales prices and attributes of the home including square footage, lot size, number of bedrooms and bathrooms, etc.
- Additional tables may be merged/joined to the main table. Frequently a merge must be performed between the estimation table and the dataset that results from the set of network-based aggregations that are described in the previous section.
- If necessary a few lines of Python/Pandas can be used to transform variables.
- Filters can be applied to remove obviously incorrect or degenerate rows of data.
- The model must actually be specified. This is usually done with Patsy, which is a highly parsimonious R-style syntax for specifying the dependent and independent variables from the dataframe generated by the steps above.

- The results must be saved to an output data store, for both estimation and simulation. For estimation, coefficients on variables must be saved and for simulation the predicted output variables are saved for use in subsequent models.

UrbanSim has been designed to take key-value pairs which specify the above set of parameters in a JSON format. An example configuration is shown in Figure 5 which gives the configuration for a linear regression on home prices and shows the parsimony of specifying a statistical estimation in this way. Each model can be specified with 10-15 key-value parameters and then models can be executed in sequence to create a simulation of the full regional real estate market (as shown in Figure 1). It is possible that this sort of framework can be expanded to be useful to the broader community of StatsModels users, but this task remains for future work.

```
{
    "model": "hedonicmodel",
    "output_table": "dset.buildings",
    "add_constant": true,
    "internalname": "buildings",
    "patsy": [
        "I(year_built < 1940)",
        "I(year_built > 2005)",
        "np.log1p(stories)",
        "ave_income",
        "poor",
        "jobs"
    ],
    "merge": {
        "table": "dset.nodes",
        "right_index": true,
        "left_on": "_node_id"
    },
    "output_varname": "nonresidential_rent",
    "table": "dset.costar",
    "table_sim": "dset.building_filter(residential=0)",
    "dep_var": "averageweightedrent",
    "segment": [
        "general_type"
    ],
    "dep_var_transform": "np.log",
    "output_transform": "np.exp"
}
```

Figure 5: A sample JSON configuration used to specify a residential sales price model.

Once models are configurable in JSON, and given that JSON is the vernacular for client-server communication on the internet, it is an incremental step to create a web service which, after specifying an HDFStore from which to read all necessary data, models can then be estimated or simulated by making http requests with a JSON model specification. A simple website has been created to read, write, and edit JSON specifications, to run sets of models in sequence, and to create charts of model results using D3 and maps of model results using Leaflet. Basic browsing of tables in the HDFStore is also available. Thus a graphical interface is underway which can be used to configure and run statistical models via a website and even to run data analysis batch jobs, and a screenshot is shown in Figure 6.

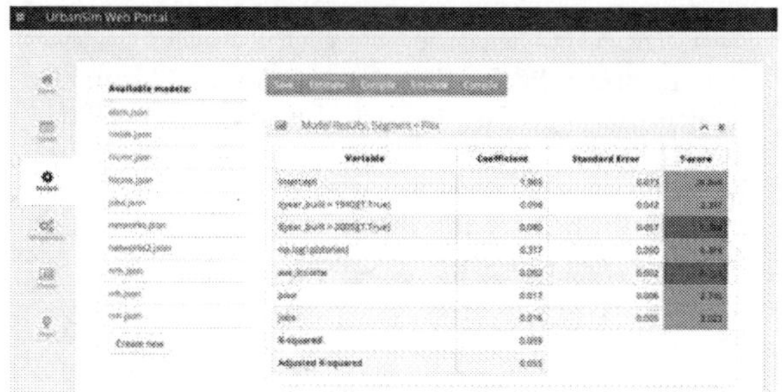

Figure 6: Screenshot of the current UrbanSim web portal.

It is worth noting that the one area that community-based scientific programming tools have been insufficient is for discrete choice multinomial (MNL) models where the number of alternatives is extremely large. For running MNL models where the number of choices is small -e.g. the choice of travel mode between auto, transit, and car -the models provided in StatsModels are sufficient. However, for location choice models in a large region, the number of alternatives can be in the thousands or even hundreds of thousands, and a probability may need to be computed for each of the alternatives.

For this case, special purpose models and model estimation code have been written which take the same basic form as the StatsModels model interface, but are not currently included in the StatsModels distribution. It is unclear at this time if these models are useful to the broader scientific community or if they are only of importance to the urban modeling community.

7 Conclusion

This paper has presented the history of UrbanSim in the context of the history of the open source projects which support it. Numerous advances have been made in Python, including Pandas and StatsModels, ease of coordinating distributed code projects using GitHub, distribution and installation of Python projects using Anaconda, and new interface technology including JSON, Angular, D3, and Leaflet. All of these advances have enabled UrbanSim to make an evolutionary leap and minimized the amount of code necessary to create a domain-specific application.

Additionally, embracing the larger open source geospatial community has allowed UrbanSim a competitive advantage over other projects which have not embraced the available tools with the same alacrity. In fact, the methodology described in this paper has now convinced the Association of MPOs to begin work on a prototype which expands the use of this methodology to activity-based travel models which are critical to the transportation planning operations in dozens of the larger regions in the United States and internationally.

The open source and community-supported nature of the core projects, including UrbanSim -and in particular the open and collaborative nature of online tools like GitHub -are garnering a positive response from proponents of open and accountable government as well as groups which support the transparency of large agent-based simulations used in governmental processes. The grassroots nature inherent in nearly all of the open source projects described here has been particularly embraced by governments, who are often reticent to support large commercial institutions and balkanized and private data formats and software tools. Clearly open source tools are well suited for applications within government, and UrbanSim is an excellent case study for progress that can be made to this end.

References

Alonso, W. (1964). Location and land use: toward a general theory of land rent. Harvard University Press.

Barbour, E. and Deakin, E. A. (2012). Smart growth planning for climate protection. Journal of the American Planning Association, 78(1):70–86.

Ben-Akiva, M., Bowman, J. L., et al. (1998). The day activity schedule approach to travel demand analysis. PhD thesis, Massachusetts Institute of Technology.

Borning, A., Ševcıková, H., and Waddell, P. (2008). A domain-specific language for urban simulation variables. In Proceedings of the 2008 international conference on Digital government research, pages 207–215. Digital Government Society of North America.

Bowman, J. L. and Ben-Akiva, M. E. (2001). Activity-based disaggregate travel demand model system with activity schedules. Transportation Research Part A: Policy and Practice, 35(1):1–28.

Brueggeman, W. B. and Fisher, J. D. (1997). Real Estate Finance and Investments. Inc.

Christaller, W. (1968). Die zentralen Orte in Suddeutschland.

de Dios Ortzar, J., Willumsen, L. G., et al. (2001). Modelling transport. Wiley.

Ewing, R. and Cervero, R. (2001). Travel and the built environment: A synthesis. Transportation Research Record: Journal of the Transportation Research Board, 1780(-1):87–114.

Foti, F. (2014). Measuring the Contribution of Walkable Amenities to Home Values and Residential Location Choices. DRAFT, University of California, Berkeley.

Geisberger, R., Sanders, P., Schultes, D., and Delling, D. (2008). Contraction hierarchies: Faster and simpler hierarchical routing in road networks. Experimental Algorithms, pages 319–333.

Grannis, R. (1998). The importance of trivial streets: Residential streets and residential segregation 1. American Journal of Sociology, 103(6):1530–1564.

Guo, J. Y. and Bhat, C. R. (2007). Operationalizing the concept of neighborhood: Application to residential location choice analysis. Journal of Transport Geography, 15(1):31–45.

Hansen, W. G. (1959). How accessibility shapes land use. Journal of the American Institute of Planners, 25(2):73–76.

Hunt, J. D. and Abraham, J. E. (2005). Design and implementation of PECAS: a generalised system for allocating economic production, exchange and consumption quantities.

Kockelman, K. and Cervero, R. (1997). Travel demand and the 3Ds: density, diversity, and design. Transportation Research Part D: Transport and Environment, 2(3):199–219.

Lee, B. H., Waddell, P., Wang, L., and Pendyala, R. M. (2010). Reexamining the influence of work and nonwork accessibility on residential location choices with a microanalytic framework. Environment and Planning A, 42(4):913–930.

Lee, D. (2009). 2009 TMA MPO modeling activity survey. Technical report, Fredricksburg Area MPO.

McFadden, D. (1978). Modelling the choice of residential location. Institute of Transportation Studies, University of California.

McKinney, W. (2012). Python for Data Analysis: Data Wrangling with Pandas, NumPy, and IPython. O'Reilly Media.

Miles, M. E., Berens, G., and Weiss, M. A. (2000). Real estate development: principles and process. Urban Land Institute Washington, DC.

Noth, M., Borning, A., and Waddell, P. (2003). An extensible, modular architecture for simulating urban development, transportation, and environmental impacts. Computers, Environment and Urban Systems, 27(2):181–203.

Openshaw, S. (1984). Ecological fallacies and the analysis of areal census data. Environment and Planning A, 16(1):17–31.

Rosen, S. (1974). Hedonic prices and implicit markets: product differentiation in pure competition. The journal of political economy, 82(1):34–55.

Tobler, W. R. (1970). A computer movie simulating urban growth in the detroit region. Economic geography, 46:234–240.

Von Thünen, J. H. and Hall, P. G. (1966). Isolated state: an English edition of Der isolierte Staat. Pergamon Press.

Waddell, P. (2000). A behavioral simulation model for metropolitan policy analysis and planning: Residential location and housing market components of Urban-Sim. Environment and Planning B: Planning and Design 2000, 27(2):247–263.

Waddell, P. (2002). UrbanSim: modeling urban development for land use, transportation, and environmental planning. Journal of the American Planning Association, 68(3):297–314.

Waddell, P., Berry, B. J. L., and Hoch, I. (1993). Residential property values in a multinodal urban area: New evidence on the implicit price of location. The Journal of Real Estate Finance and Economics, 7(2):117–141.

Waddell, P., Ševcıková, H., Socha, D., Miller, E., and Nagel, K. (2005). Opus: An open platform for urban simulation. In Computers in Urban Planning and Urban Management Conference, London.

Wegener, M. (2004). Overview of land use transport models. Handbook of transport geography and spatial systems, 5:127–146.

Adding Phylogenies to QGIS and Lifemapper

for Evolutionary Studies of Species Diversity

by Jeffery A. Cavner, Aimee M. Stewart, Charles J. Grady, James H. Beach

University of Kansas (USA). `jcavner@ku.edu`

Abstract

Phylogenetic data from the "Tree of Life" have explicit spatial and temporal components when paired with species distribution and ecological data for testing contributions to biological community assembly at different geographic scales of species interaction. Important questions in biology about the degree of niche suitability and whether the history of a community's assembly for an area can affect whether the species in a community are more or less phylogenetically related can be answered using several different spatially-filtered measures of phylogenetic diversity. Phylogenetic analyses which support the description of ecological processes are usually achieved in a handful of software libraries that are narrowly focused on a single set of tasks. Very few applications scale to large datasets and most do not have an explicit spatial component without relying on external visualization packages. This prompted us to explore bringing phylogenetic data into an open-source GIS environment. The Lifemapper Macroecology/Range & Diversity QGIS plug-in is a custom plug-in which we use to calculate and map biodiversity indices that describe range-diversity relationships derived from large multi-species datasets. We describe extensions to that plug-in which expand the Lifemapper set of ecological tools to link phylogenies to spatially-derived 'diversity field' statistics that describe the phylogenetic composition of natural communities.

Keywords: QGIS, WPS, Distributed Computing, Biogeography, Range and Diversity, Lifemapper, Macroecology, Phylogenetics.

1. Background

Community phylogenetics, the focus on how species relatedness and species traits are associated with how evolution extends into ecological processes and spatial patterns, and biogeography or meta-community ecology, largely focused on the spatial regulation of species distributions, should assay the spatial variation of phylogenies by mapping phylogenetic community values across space and time at different scales using advances in GIS techniques. One such approach would to be bring phylogenetic data into a GIS environment. We have begun to develop such an approach as an addition to the Lifemapper project (www.lifemapper.org) in a Lifemapper Range & Diversity (LmRAD) QGIS plug-in (Cavner et al. 2014) that provides phylogenetic visualization and analysis tools for spatially linked range-diversity relationships derived from presence-absence matrices (PAMs). We developed the tool also hoping to expand it to include historical biogeography meta-community analyses and community assembly analyses focused on phylogenetic-diversity area relationships where analysis across geographic scale leads some of the most important questions in biodiversity.

The LmRAD QGIS plug-in creates, maps and analyzes presence-absence matrices or PAMs, one of the core data structures for macroecological research. It links the resulting data to phylogenetic and spatial views of a set of range-diversity statistics derived from the PAM. The PAM or incidence matrix is a 2-dimensional Boolean matrix constructed from a spatially defined grid of regular polygons where the presence or absence of each species of hundreds or thousands of species are recorded for each cell. One axes of the matrix represents species and the orthogonal axis represents geographic localities described by the regular polygons. Each geographic site is coded for the presence (1) or absence (0) of each species. It summarizes the two fundamental units of biogeography, the distributional range of a species (both their position and size, range size simply equals the total of the species axes across sites) and the species diversity of sites or the number of different species in each site as summarized by site axes totals.

Several mathematical and biological relationships obtain across the PAM that link spatially derived statistics with species based statistics. Of interest for phylogenetic relationships are the species based statistics calculated from the PAM that measure the "diversity field" of a species (Arita et al. 2008). The diversity field is the set of diversity values of sites in which a species occurs. For example, the diversity field volume, i.e. the summation of those species diversity values within a species' range divided by the range size of the species allows us to calculate the average species diversity within the range of that species. We represent that volume as a proportion of the total number of species in the study area. Including the total area of the study area allows us to illustrate the proportion of the sites in which two species co-occur. The average association of a species with all of the species in the study area allows us to illustrate that there is an inverse relationship between the proportional range of a species and the difference between the mean proportional diversity within its range and the average proportional diversity in the study area (Arita et al. 2008). The mathematical reciprocal of the average proportional diversity of the study area is a well-studied measure of species turnover called Whittaker's beta diversity. It is a measure of the ratio between the overall diversity of the study area and the average local diversity (Arita et al. 2008). There are closely associated beta measures of diversity for several different types of diversity. Different approaches to species diversity such as phylogenetic diversity – the degree of relatedness of species in a community based on their evolutionary history – abundance and ecosystem function measures of diversity all can be decomposed into measures of local and regional diversity ratios that are highly dependent on scale.

Analyzing the diversity field within the range of a species is equivalent to studying it's covariance with all the species in a study, i.e. the degree of association of species within their ranges. We plot this association in QGIS through the plug-in in a "range-diversity" plot. Curves on the plot for species follow a line defined by the inverse relationship between the range of a species and the difference between the two diversity statistics. When plotting the species in this way, species with equal degrees of association with one another arrange themselves along lines of isocovariance. The Lifemapper plug-in allows the user to "brush" data points along those curves in the interactive range-diversity plot which selects the individual species in the linked data space for the phylogenetic tree. In this way the spatially derived statistics for diversity from the PAM can be compared to the degree of phylogenetic relatedness within species communities.

The plug-in accomplishes this by using QGIS as a WPS client to Lifemapper web services (Stewart et al. 2014) and by using JavaScript based visualization technologies for large phylogenetic trees within the plug-in. Macroecology algorithms are exposed as Open Geospatial Consortium (OGC) Web Processing Services (WPS) (Open Geospatial Consortium, Inc. 2007b) so that larger distributed computing environments can be brought to bear on large datasets. The Lifemapper web services are organized as two modules, LmSDM, and LmRAD. The LmSDM module uses RESTful and OGC specifications to build species distribution models based on the predicted niche for a species using climate and species occurrence data. The LmRAD (Range and Diversity) is a multi-species platform for PAM based range and diversity calculations. Both modules can be accessed through the plug-in, and outputs from LmSDM can be piped into LmRAD as species inputs to PAMs. This paper will focus on the range and diversity capabilities of the plug-in and how the spatial component to phylogenetic data recently added to the plug-in can be used with the biodiversity indices calculated from the PAM and areas where phylogenetic data can be used to explore other types of diversity measures for species communities. This paper will begin by outlining use cases and common threads that connect them and how we have begun to address them with a focus on new interface capabilities for phylogenetic data and linked data spaces. Next we will describe how the Lifemapper plug-in and it's supporting web services were designed to take advantage of a client-server architecture in order to be able to use geographic processing standards on large datasets. This is followed by a comparison of related software with a focus on phylogenetic algorithms and scripts with a spatial component. We end by discussing findings, and future directions for the Lifemapper plug-in.

2. Use Cases and Capabilities

2.1 Range and Diversity Plots and Maps with Phylogenetic Trees

Phylogenetic based ecology is a growing field. Its practice both at small scales and larger biogeographic scales – it goes under several names: phylogeography, ecophylogenetics, or phylogenetic community ecology – share two obvious constraints for incorporating phylogenetic data into ecology research. First, many ecophylogenetic methods are not available as open-source software packages, and are therefore not extensible or customizable, and second; the tools are scattered across specialty software each with their own learning curve and with unique data formats (Kembel et al. 2010). When

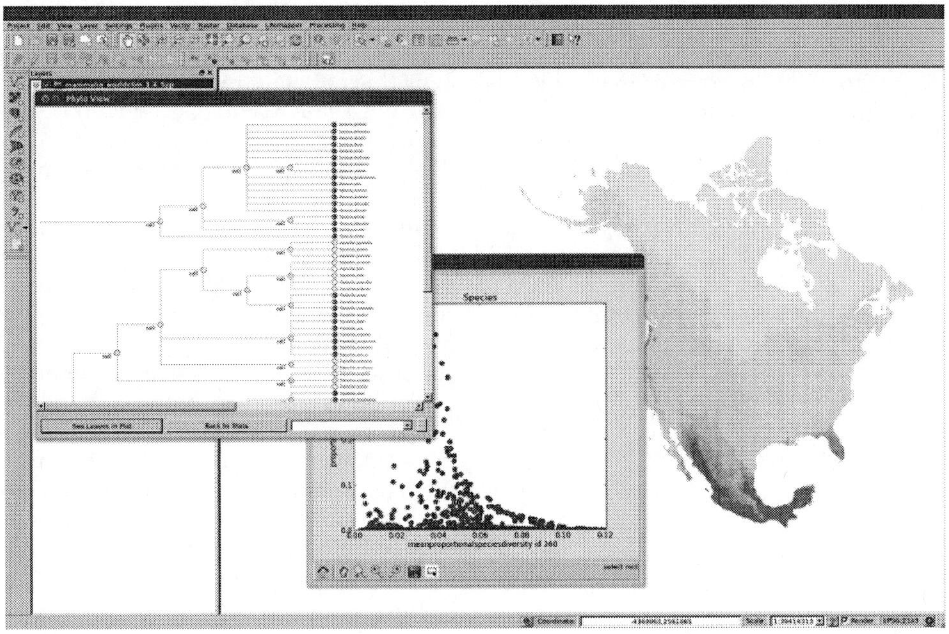

Figure 1: Lifemapper Range-Diversity Plug-in in QGIS with Range-Diversity Plot and Tree View.

we extended Lifemapper to include the multi-species range and diversity experiments on PAMs in the LmRAD module we encountered a third constraint. Untangling species associations at large scales remains an important research question (Arita et al 2008) and these scales require large taxa with numerous species in single experiments, e.g. approximately 3000 + species of birds in South America. Each of these species represents some type of geographical map which have to be intersected with large numbers of polygons resulting in PAM matrices of several million elements. These models must also be permuted, i.e. randomized to perform null model hypothesis testing, resulting in a large number of unwieldy data structures. Additionally these studies need to be done at several different spatial scales and across time. Totaling all of these dimensions one can readily see that this presents serious computational challenges. These factors informed the design of LmRAD as a module to the existing Lifemapper computational platform which uses co-located distributed high-through-put compute resources to calculate large multi-part ecological modeling jobs, and a thick client front-end in QGIS to those services.

Important biological relationships are expressed in the PAM as associations between species diversity in communities and the range size of those species. Two important correlations are between the species diversity of sites and mean range sizes of species occurring in the site; and between the range sizes of species and the mean species diversity within those ranges. The set of range statistics for species within a site are described by a 'dispersion field'. The analogue to that measure for species is the 'diversity field' which quantifies the species diversity of sites in which a species occurs (Arita et al. 2008). Range-diversity plots are produced in the Lifemapper plug-in that summarize these fields as indexes of site similarity and the degree of association of species adding to our knowledge of species communities. Data points in the plots are constrained by the association of species and site similarity and the proportional fill of the matrix, i.e. the ratio of presences of species to their absence. The plug-in allows a researcher to build several models across scale, experiment with fill, extent and resolution. The dispersion field and diversity field measures in the range-diversity plots are interactive and allow multiple pane selection. Visualization and data exploration are presented in both geographic and phylogenetic data spaces. The dispersion field statistics are viewed in an interactive "by-sites" range-diversity plot and are linked geographically to a map of the range statistics attached to the input grid for the PAM so that they can be overlaid with other geographic data. For species associations within communities the data derived from the PAM are depicted in an interactive "by-species" range-diversity plot for the diversity field and are linked to a dendrogram that represents their phylogenetic relationship. All of the data spaces allow for 'brushing' of datasets by species or location across tree data space and geographic data space. Selecting species in the phylogenetic tree viewer select those species in the "by-species" range-diversity plot. In this way the trees can act as a data exploration tool against the diversity indices derived from the PAM providing insight into the phylogenetic composition of communities where species co-occur. (see figure 1.)

2.2. Across Space and Time: Scale Considerations

The indices currently calculated through the plug-in, including ecology staples, such as beta diversity, along with measures of nestedness – the degree to which diversity loss occurs by species, leaving isolated "islands" of diversity – are all effected by scale. The degree to which these indices are effected by scale and the mechanisms involved are important research questions (Arita et al. 2008, Lira-Noriega et al. 2007). Most analyses of scaling effects on diversity have been based on coarse input grids. For example Hawkins et al. (2003) based a diversity study comparing the effect of scale using 85 datasets with resolutions ranging from 103 to 105 km2 (Lira-Noriega et al. 2007). Lira et al. performed a study with finer PAM resolutions starting at 11.4 km2 and incrementally climbing to 2.93 x 103 km2 for an area of ˜ 138,200 km2. The Lifemapper plug-in has been used to construct PAMs for much larger areas, ˜ 24,709,000 km2 with slightly larger cell resolutions of 100 km2, but with the recent additions of data parallelization and portable instances of Lifemapper we expect to be able to produce PAMs with cell resolutions lower than 1/32o for the globe. We can also currently test scale related hypotheses about range size and diversity such as predictions that for the same kind of organism, organized by taxa, and their ability to disperse across the landscape, stronger negative correlations between range size and diversity should exist the greater the scale. Several questions that relate to spatial scale can also be asked of phylogenetic-diversity area relationships, and the extent to which speciation and adaptation contribute to community assembly with the incorporation of phylogenetic tree data into the plug-in.

Because biogeographers are increasingly interested in methods in phylogeography and community assembly, research questions addressed by both species richness based diversity measures, phylogenetic diversity and functional diversity need to benefit from relative findings and work together to complement one another (Cianciarus 2011). A common thread connecting different concepts of diversity are questions about the evolutionary and biogeographical history of a species and how temporal and spatial scales affect the evolutionary relatedness of species in a habitat and the degree that those assemblages are consistent with environmental filtering or competitive interaction (Emerson and Gillespie 2008). The species composition of natural communities is tied to questions of range contraction and local extirpation of species in relation to niche processes like climate change. The Lifemapper/QGIS plug-in allows the user to build PAMs that describe range and diversity relationships across time in relation to climate change by using predicted eco-niches based on climate scenarios, derived from LmSDM, as inputs to future PAMs.

Phylogenetic data has both spatial and temporal components. Patterns of co-occurrence of species in a spatially defined community is effected over different time and spatial scales by the similarity, and distance of other habitats, the degree that niches are filled with current inhabitants and the relative time available for colonization or adaptation (Emerson and Gillespie 2008). Patterns of community structure and co-occurrence of species can be summarized by two related statistics derived from phylogenies for a geographic area, phylogenetic clustering, and phylogenetic over-dispersion/evenness. Phylogenetic clustering occurs when co-occurring species are more closely related than can be expected by chance. Phylogenetic over-dispersion/evenness occurs when co-occurring species are more distantly related than can be expected by chance. With the tree viewer these phenomena are easily discernible for small trees with species selected that co-occur within a community. Both of these measures will need to be quantified for larger trees and both require that they be tested against null models generated from the tree and its spatial components. Lifemapper currently implements some very efficient bit-wise operations for randomizing null models from the PAM. To permute the tree data, we will in the future build out the architecture for encoding the tree topology from large phylogenies into matrices that will use similar methods for randomization.

Clade based analyses of traits related to niche occupancy helps us to understand the relative importance of environmental filtering. Using cross scale comparisons in the plug-in with the phylogenetic trees could help to tease out effects of both temporal and spatial scale. Larger extents within an LmRAD experiment should show phylogenetic clustering due to environmental filters, and local areas which will naturally contain subsets of the same taxa used in the experiment should show local overdispersion due to competitive interaction. For temporal scale, range and diversity measures from time-stepped PAMs achievable with the recent acquisition of paleontological climate layers, and the future climate scenario data currently in the plug-in, should allow us, with the use of the trees be able to look at colonization dynamics, and how over-dispersion and cladogenesis become more important over time for isolated niches and how species new to a habitat over large time frames, e.g. island migration, show shared common traits pre-adapted to a habitat (Emerson and Gillespie 2008).

3. Design and Architecture

3.1 Lifemapper Distributed Computational Services

The Lifemapper Range and Diversity (LmRAD) module is an analysis suite that extends the current Lifemapper (www.lifemapper.org) platform allowing us to leverage the computational power of distributed computing environments to execute the range-diversity analyses as distributed algorithms. The algorithms are exposed as Open Geospatial Consortium Web Processing Services (WPS) (Open Geospatial Consortium, Inc. 2007b), and RESTful web-services for simple data retrieval and viewing. The Lifemapper infrastructure is composed of a central management component, LmDbServer, which manages data and analysis operations with a "data pipeline" written in Python (www.python.org) and a PostgreSQL/PostGIS database; multiple instances of LmCompute that can be co-located across institutions, currently deployed at compute clusters at University of Kansas, University of Florida, and San Diego Supercomputer Center; a continuously updated species model and species occurrence set archive based on museum data for species from the Global Biodiversity Information Facility (GBIF); and LmWebServer which manages all communications between the components and client applications. (see Figure 2.) LmRAD specifically is a distributed mulit-species modeling module within this system with custom algorithms for working with presence-absence data, including matrix definition, construction, calculation, randomization for null models and preparation of visualization outputs, trees and maps.

As a job based infrastructure, LmRAD and LmSDM algorithms are environmentally agnostic and are portable across compute environments through instances of LmCompute that are deployable in several types of distributed compute environments. LmCompute is a pluggable, configurable, open source client that abstracts the details of the compute job away from the physical system. LmWebServer contains a Job Server tier that feeds jobs to any compute environment that can sponsor an instance of LmCompute. LmCompute is also generalizable, since LmCompute only interacts with the physical system through a mediator designed along the mediator and facade design patterns (Gamma et al. 1994) the compute plug-in expects just a few stock functions. A "request job" method call might just as easily get a local XML job definition or pull a job from the Lifemapper Job Server. An instance of LmCompute can use a job response to instantiate a Job Runner object and retrieve inputs to the methods requested. Each of these computational tasks or group of related tasks is a compute plug-in based on the template method and strategy design pattern (Gamma et al. 1994). The compute plug-in is wrapped in a "runner" class that depending on its run method can execute an external application or run custom algorithms like LmRAD algorithms. A compute plug-in receives its jobs through a job controller that acts as a hub for producing job outputs. Using the factory method pattern and command pattern (Gamma et al. 1994), the controller sits in front of a compute environment, requests data inputs for a job, and determines through Python "duck typing" which compute plug-in is appropriate for the computation. The pipeline and LmDbServer are responsible for presenting jobs to the Job Server on LmWebserver and moving jobs through the system. At different stages in a LmRAD experiment dependencies and statuses are updated by LmCompute which posts back to the Job Server during the process. LmRAD PAM operations specifically have been parallelized across processors on any compute environment that receives a PAM job. Data products for large PAMs at high resolutions (10 km) with upwards of 800 species can be constructed and analyzed in this way with reasonable response times. Results from the experiment are then posted back to the Job Server from the compute environment and are written to the database and file system shared by the LmDbServer and LmWebserver.

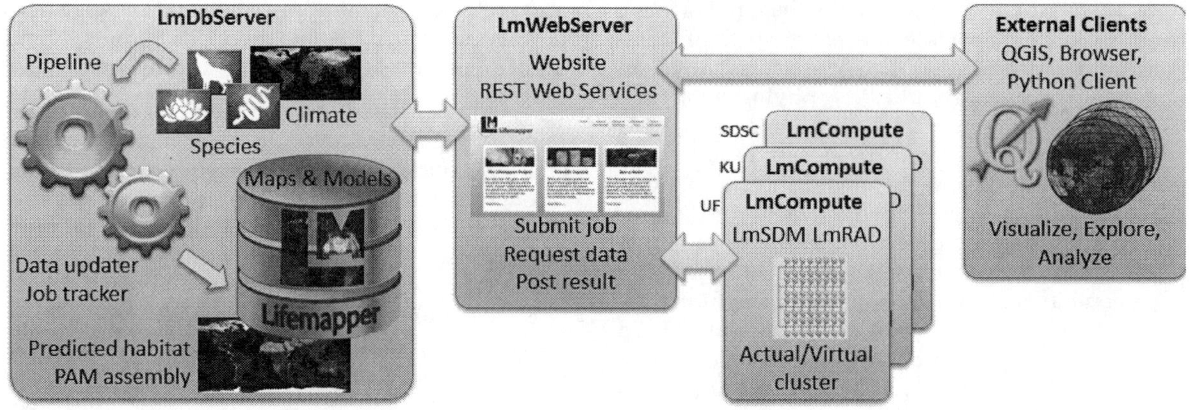

Figure 2: Lifemapper Components.

Data parallelization across multi-core architectures in each of the environments hosting LmCompute help to speed the PAM matrix construction, which uses a combination of Rtree (https://pypi.python.org/pypi/Rtree/) and matplotlib's nxutils and GDAL (GDAL 2013) for vector and raster based intersections, respectively. Calculations on the matrices use NumPy (Jones 2001) built with the Basic Linear Algebra Subprograms (BLAS). Permutations on the PAM matrices use methods that are specific to binary matrices, where row and column totals can both be kept intact while changing the mix of species in sites and the range size of each species. Data parallelization is not suited to these computations since the entire matrix needs to be taken into account. But since several hundred permutations may be required per experiment, the current job based parallelization across compute nodes works well for permuting several hundred matrices. Another method in LmRAD for permuting the matrix is perfectly suited for both types of parallelization. It uses a dye dispersion algorithm which is a 2-Dimensional geometric-constraints model that assumes range continuity (Jetz and Rahbek 2001). Since range allocations are reassembled individually for each species, those data can be split across cores on a single machine or across nodes.

3.2 QGIS Plug-in, WPS Client, JavaScript, Plots, Visualization

The computational constraints of operating against large matrices in current desktop software informed the design of LmRAD as a client-server architecture using web-services to off-load the heavy lifting for PAM operations to remote compute environments with the use of a thick client inside a feature-rich open source GIS environment. The Lifemapper plug-in for QGIS allows QGIS to operate as a web service client to the LmSDM services and as a WPS client to the LmRAD analyses, edit and submit data, parameterize inputs and request computations with the added feature of being able to pull down statistical results, geospatial outputs and work with phylogenetic tree data. It is able to do this by interacting with a multi-platform Python client library that abstracts the communication layer away from the user. The Lm client library can also be decoupled from the client so that developers can use it to program a variety of standards based clients. The added benefit of using the library within the QGIS plug-in is that all LmRAD functionality for dealing with PAM operations are wrapped in easy to use, point and click operations with results automatically downloaded to a managed workspace and presented in QGIS.

Viewing the phylogenetic trees required that a highly interactive and lightweight interface be built in the plug-in without library dependencies. Rather than deal with heavy Qt (http://qt-project.org/) solutions for graphics we decided to leverage recent advances in web based standards for visualization of phylogenetic trees in the QGIS plug-in using a document driven JavaScript framework. Tree data from the phylogenetic community can take any one of several forms, phyloXML (www.phyloxml.org), Newick (http://bit.ly/1n6ELcZ), Nexus (Maddison et al. 1997), and NeXML (http://nexml.org). All of these formats are easily translated into JSON which maps into Python dictionaries and works well with web standards based solutions for visualizations based on JavaScript. Additionally tree providers, like Open Tree of Life (http://blog.opentreeoflife.org/) are developing NexSON, a badgerfish convention JSON translation of Newick as a data document for transport from web-services that provide trees served from graph databases. Data like these are perfect for producing a scene graph, can be made available from web-services, are easily transported back and forth from LmCompute for analysis and can be used directly in a document driven visualization framework.

The tree viewer presents the phylogenetic data as interactive SVG built dynamically from incrementally loaded JSON data. This is made possible with the JavaScript library D3.js (Data Driven Documents) (http://d3js.org/). D3 allows the JSON document to be dynamically bound to the Document Object Model so that data-driven transformations can be applied to the document with smooth transitions and fluid interaction. The data are directly mapped to visual elements in the DOM without an internal or intermediate representation or abstraction of the DOM. The document is the scene graph. This allows for much better performance since the focus is on transformation of the document (Bostock et al. 2011). Selections against the DOM are declarative in a functional programming style with predicates from the W3C Selectors API similar to jQuery allowing CSS properties to be specified as functions. Incoming data can create new nodes in the DOM, and outgoing data can remove nodes using Enter and Exit selections. This is especially useful when navigating large trees, since the large number of nodes and edges for large phylogenies have in the past been hurdles for visualizing tree data in a way that is responsive to user interaction conditioned to fast response times.

The D3 based interactive tree is rendered in the plug-in through a Qt dialog using QtWebKit. Communication between the tree and the rest of the plug-in is effected by QtWebKit Bridge. The bridge allows the JavaScript and PyQt objects to communicate with one another. The tree viewer is linked to the interactive range-diversity plots in matplotlib (Hunter 2007) by simple PyQt signals and slots. A similar method connects the range-diversity plots for site-based statistics to the maps in QGIS based on the PAM. Using JavaScript in PyQt dialogs for QGIS allowed us to achieve fluid visual representations of trees for large clades, e.g. one tree used in testing is the entire phylogeny for the Phylum Mollusca with over 85,000 nodes.

4. Comparison of Approaches

Several phylogenetic analysis software implementations exist, the number is too daunting to recount them all here and most are implemented in R scripts and free but not necessarily open C++ software. Very few integrated systems exist that address biogeography, species communities, ecological niche and phylogeny. With the growth in phylogenetic data, web-based solutions for viewing trees are popular, but those concentrate on data already analyzed for specific taxa and tend to illustrate simple clade-area relationships. Challenges for both analyzing and exploring large phylogenies exist both on the computation side and the visualization side. We mention some very powerful approaches that contain a spatial component in relation to phylogenetic analysis and compare them to our tool which aims at bringing phylogenetic data into a GIS based tool that is sustainable and extensible using an analysis, that until now has not been systematized, using PAMs and their inherent range-diversity relationships

GeoSSE (Geographic State Speciation and Extinction, Goldberg et al. 2011) is a geographic range/-phylogeny model. GeoSSE is an an extension of the BiSSE (binary state speciation-extinction) model that allows tests for relationships between speciation or extinction and geographic range. GeoSSE is a method for analyzing the reciprocal influence of character traits and speciation/extinction, where character states are defined by spatial distributions. Transitions between states are parametrized in terms of range expansion through dispersal and range contraction through local extirpation. The model has the liability of requiring fairly large phylogenies with one or two hundred species at the leaf nodes as a minimum. The increasing availability of larger trees shouldn't make this much of a problem in the future, but may potentially also require computational solutions addressed by a distributed or parallel implementation.

Picante (Kembel et al. 2010) is a comprehensive R package for calculating phylogenetic diversity of ecological communities. It contains functions for both local or alpha phylogenetic diversity and beta phylogenetic diversity. Local community diversity indexes include Faith's phylogenetic diversity (PD) (Faith 1992), taxonomic distinctness indexes, mean pairwise phylogenetic distance (MPD) and mean nearest taxon distance (MNTD) within communities. Clustering and evenness are represented by several measures calculated in Picante. Beta phylogenetic diversity is also addressed with MPD and MNTD between communities, Sorenson index and the UniFrac phylogenetic distance metric. Picante also has robust null model capability, performing numerous permutation procedures. Ecological correlation is also included with species-environmental regressions. Picante would be an extremely powerful addition to a workflow involving large matrices using parallel methods in R or a framework like Lifemapper. Picante's methods are staples and starting points for numerous different analyses that could be performed in QGIS, benefiting from an explicit spatial component especially in regards to its ecological links to phylogenetic statistics.

Landis, Matzke, Moore, Huelsenbeck (2013), recognize that the main constraints on using models to describe the geographic evolution of species ranges as processes of dispersal and extinction is the computational limit on the number of areas that can specified. Where Lifemapper choose to leverage distributed computational resources to solve similar scale problems for large numbers of sites the Landis et al. method uses a Bayesian approach for inferring biogeographic history that allows more realistic problems involving large numbers of geographic sites implemented in BayArea, a free C++ command-line program that uses PAMs and phylogenetic data in the Newick format as inputs. Its outputs can be visualized as tree/map animations in an external JavaScript web service for filtering phylogenetic reconstructions and mapping them.

Biodiverse (Laffan, Lubarsky, Rosauer 2010), an open-source project similar to the Lifemapper plugin, provides linked visualization across different data spaces. Biodiverse links species distributions in geographic, phylogenetic, taxonomic and matrix space. One advantage of Biodiverse similar to Lifemapper is that scale comparison are achieved through a window analysis for endemism, phylogenetic diversity, and beta diversity. By varying the size of the windows one can start to understand the effects of scale on those statistics. Currently the Lifemapper plugin uses a multi-grid approach where several subsets at different cell resolution can be built out within the same experiment allowing comparisons across scale for the range and diversity statistics including beta diversity.

5. Future Directions and Conclusion

5.1 Incorporation of R for ad-hoc phylogenetic diversity-area measures against a PAM archive

The Lifemapper Project is exploring mapping it's algorithms into a MapReduce paradigm using an Apache Hadoop-based Architecture (HBA) and software-defined systems (SDS) and Multiple-Domain Distribution/Replication (MDD) of Lifemapper itself as part of a push for investment in sustainable biodiversity cyberinfrastructre. Allowing Lifemapper to live at other institutions through MDD will allow platform owners to define the types of analyses supported by Lifemapper meeting an ever growing need for more flexible and ad-hoc algorithm deployment. Researchers in the areas of bioinformatics that Lifemapper supports live in a world dominated by R scripting. Parallelizing R for Hadoop, using one of several well established methods for this, like R+Hadoop or RHIPE may allow us to calculate larger jobs in a finer grained manner, allowing code reuse, and uncoupling analyses from siloed stacks in Python on LmCompute.

A useful application of this would be the calculation of phylogenetic-diversity, overdispersion/evenness and clustering for user defined subsets of a PAM archive or Global PAM (GPAM). With the GPAM, PAM construction could be pipelined and a continuously updated PAM archive for all the world's terrestrial species from GBIF could be sub-setted, both taxonomically and spatially, by a user for on-demand data needs. Phylogenetic trees would have to be resolved from tree provider services, now coming on-line, for the species in the PAM, and Lifemapper services could enable those data through a phylo-to-matrix module, that would abstract the phylogenetic topology into a series of matrices and provide permutations of the phylogenetic data for hypotheses testing. These products would have several over-linking uses across different types of analyses. Such a PAM archive and its computational architecture for distributing matrix math across compute resources could also support the quantitative evaluation of the joint effects of historic biogeographic events to test whether different species are more or less constrained by past biogeographic events. A meta-community analysis like this is outlined by Leibold et al. 2010, where the degree of contingent historical constraint is compared to environmental suitability across a phylogeny using correlation matrices derived from several types of data. The authors of this method point to the need for addressing issues of range shifts and phylogenetic adaptation in meta-communities across several clades requiring extensive phylogenetic information (Leibold et al. 2010). Adding more robust phylogenetic based analyses to models in Lifemapper in combination with the niche models in its archive would be a valuable resource for such an analysis.

5.2 Conclusion

We have summarized an on-going effort to incorporate phylogenetic data into a flexible computational platform for multi-species range and diversity modeling in order to bring a more complete history of the diversity patterns of species' communities into focus. Concentrating on range-diversity relationships and a species 'diversity field' derived from calculations on large matrices presented to a thick GIS client in QGIS as web-services allowed us to build a set of robust tools that leveraged open-software, and exposed those analyses to a larger audience, enabling transformative new science. The addition of phylogenetic data to the range-diversity plots and maps allows a user to explore community assembly of species habitats and answer questions about dispersal, competition and adaptation to the environment.

With the explosion of data across all areas of ecology and especially in the phylogenetic community, the need for scalable software solutions for dealing with computationally intensive calculations on large datasets is increasingly clear. Common to most of the methods discussed for analyzing phylogenies is the wish to combine them with environmental data and species range data. Macroecology and biogeography are becoming more cross-disciplinary and are incorporating more methods from community phylogenetics. As this happens phylogenetic datasets will need to reach across more of the tree of life. Spatially they will become biogeographical in scale requiring that researchers have access to computational resources not easily accessible to non-computer specialists. A set of phylogenetic community ecology algorithms that leverage those resources through a suite of web services with a thick client should be designed for maximum flexibility allowing code reuse, and definable by the end user freeing the researcher to concentrate on formulating and testing hypotheses in order to be able to describe the earth's diversity and answer important questions about the fate of our planet's health. Lifemapper is a computational platform that answers some of these challenges, it has implemented a suite of range-diversity statistics never before formalized in relation to phylogenetic data, with a unique interface which scales to large phylogenetic trees, embedded within a rich spatial GIS environment.

Acknowledgements: Authors were supported by NSF/BIO/AVAToL Award #1208472. We are grateful to our colleagues and collaborators, Jorge Soberon, Andres Lira-Noreiga and Rafe Brown.

References

Arita, H. T., Christen, J. A., Rodriguez, P., & Soberón, J. (2008). 'Species diversity and distribution in presence-absence matrices: mathematical relationships and biological implications.' The American Naturalist, 172(4), 519-532.

Bostock, M., Ogievetsky, V., & Heer, J. (2011). 'Dě data-driven documents.' Visualization and Computer Graphics, IEEE Transactions on, 17(12), 2301-2309.

Cavner, J.A., Beach, J.H. Stewart, A.M. Grady, C.J (2014) Lifemapper Macroecology Range and Diversity Tools v. 2.0.1 [QGIS plugin, Computer Software], Lawrence, KS: University of Kansas Biodiversity Institute. Available from http://plugins.qgis.org/plugins/lifemapperTools/

Cavner, J. A., Stewart, A. M., Grady, C. J., & Beach, J. H. (2012). 'An innovative Web Processing Services based GIS architecture for global biogeographic analyses of species distributions'. OSGeo Journal, 10(1), 11.

Cianciaruso, M. V. (2011). 'update: Beyond taxonomical space: large-scale ecology meets functional and phylogenetic diversity.' Frontiers of Biogeography, 3(3).

Emerson, B. C., & Gillespie, R. G. (2008). 'Phylogenetic analysis of community assembly and structure over space and time.' Trends in Ecology & Evolution, 23(11), 619-630.

Faith, D. P. (1992). 'Conservation evaluation and phylogenetic diversity'. Biological Conservation, 61(1), 1-10.

Gamma, E., Helm, R., Johnson, R., & Vlissides, J. (1994). Design patterns: elements of reusable object-oriented software. Pearson Education.

GDAL (2013) Geospatial Data Abstraction Library: version 1.10.1 Open Source Geospatial Foundation, http://gdal.osgeo.org

Goldberg, E. E., Lancaster, L. T., & Ree, R. H. (2011). 'Phylogenetic inference of reciprocal effects between geographic range evolution and diversification.' Systematic Biology, 60(4), 451-465.

Jetz, W. and Rahbek, C. (2001) 'Geometric constraints explain much of the species richness pattern in African birds.' Proc. Nat. Acad. Sci. USA 98:5661–5666

Jones, E. (2001) SciPy: Open Source Scientific Tools for Python Url: http://www.scipy.org/

Kembel, S. W., Cowan, P. D., Helmus, M. R., Cornwell, W. K., Morlon, H., Ackerly, D. D., ... & Webb, C. O. (2010). Picante: R tools for integrating phylogenies and ecology. Bioinformatics, 26(11), 1463-1464.

Laffan, S. W., Lubarsky, E., & Rosauer, D. F. (2010). 'Biodiverse, a tool for the spatial analysis of biological and related diversity. ' Ecography, 33(4), 643–647.

Landis, M. J., Matzke, N. J., Moore, B. R., & Huelsenbeck, J. P. (2013). 'Bayesian analysis of biogeography when the number of areas is large.' Systematic biology, 62(6), 789-804.

Leibold, M. A., Economo, E. P., & Peres-Neto, P. (2010). Metacommunity phylogenetics: separating the roles of environmental filters and historical biogeography. Ecology Letters, 13(10), 1290-1299.

Maddison, D. R., Swofford, D. L., & Maddison, W. P. (1997). NEXUS: an extensible file format for systematic information. Systematic Biology, 46(4), 590-621.

OGC 2007b OpenGIS Web Processing Service, Version 1.0.0. Wayland, MA, OGC Document No 05-007r7

Scheiner, S. M. (2012). ' A metric of biodiversity that integrates abundance, phylogeny, and function.' Oikos, 121(8), 1191-1202.

Stewart, A.M., Beach, J.H., Grady, C.J., Cavner, J.A. (2014) Lifemapper [Computational platform services for species distribution modeling and continental-scale biodiversity pattern analyses] Web: www.lifemapper.org

Evaluation of Web Processing Service Frameworks

by M. Ebrahim Poorazizi and Andrew J.S. Hunter

University of Calgary (Canada). mepooraz@ucalgary.ca

Abstract

As geoprocessing on the web has matured in recent years, an increasing number of geoprocessing services and functionality are becoming available in the form of online Web Processing Services (WPS). Consequently, the quality of such geoprocessing services is of importance to ensure that WPS instances fulfill users' expectations. In this paper, we illustrate, and discuss initial results from a quantitative analysis of the performance of WPS servers. To do so, we used two test scenarios to measure response time, response size, throughput, and failure rate of five WPS servers including 52° North, Deegree, GeoServer, PyWPS, and Zoo. We also assess each WPS server in terms of qualitative metrics such as software architecture, perceived ease of use, flexibility of deployment, and quality of documentation. A case study addressing accessibility assessment is used to evaluate the relative advantages and disadvantages of each implementation, and point to challenges experienced while working with these WPS servers.

Keywords: OGC WPS; Geoprocessing; Performance Evaluation; Benchmark.

1 Introduction

With the development of geospatial services, web-based GIS (Geographic Information Systems) have progressed towards a service-oriented paradigm (Mayer, Stollberg, & Zipf, 2009). Today, spatial services can be used to effectively support common tasks undertaken by spatial information users, for example, discovery and access to, process of, or visualization of spatial data. Catalogue Services for the Web (CSW), Web Feature Services (WFS), Web Coverage Services (WCS), Web Mapping Services (WMS), and WPS are common services defined by the OWS (Open Geospatial Consortium Web Service) initiative. A CSW provides the ability to publish and search collections of descriptive information (metadata) (Solntseff & Yezerski, 1974) for spatial data and services (Nebert, Whiteside, & Vretanos, 2007). A WFS is the main geospatial service for publishing vector spatial data, generally encoded using Geography Markup Language (GML) (Vretanos, 2002). A WCS defines a standard interface and operations that enable interoperable access to spatial coverage (Spatial information representing space/time-varying phenomena) datasets (Evans, 2003). A WMS delivers visualizations of data and, unlike WFS and WCS, does not deliver the data directly (de La Beaujardiere, 2004).

In the context of processing services, the Open Geospatial Consortium (OGC) has standardized the WPS interface for publishing of spatial processes, the discovery of, and binding to, those processes by users (Schut, 2007). A spatial process may include algorithms, calculations, or various kinds of models, which are exposed as a service instance, and operating on spatial data. A WPS, thus, can be used to design and develop a wide variety of GIS functionalities, and be made available to users across a network, as well as provide access to previously defined functions, calculations, or computational models.

With the emergence of geoprocessing on the web, the WPS specification and its (application) profiles have been applied to a wide array of use cases, from accessibility assessment (Steiniger, Poorazizi, & Hunter, 2013) to ecological modeling (Dubois, Schulz, Skøien, Bastin, & Peedell, 2013). The increasing use of WPS instances has also raised pertinent quality concerns — users/developers are likely to be concerned about the Quality of Service (QoS) attributes such as performance, reliability, and security.

The performance of a particular WPS is often of importance to users, arguably the most important, when evaluating the QoS of a specific service. Moreover, performance has a direct effect on other QoS attributes; for example, poor performance will affect reliability, scalability, capacity, accuracy, accessibility, and availability (Cibulka, 2013).

A developer's concerns, during designing and development of a WPS, are often twofold. As noted, from a quantitative perspective, performance is one of the key principles that can ensure both user and application developer satisfaction. From a qualitative point of view, quality concerns such as software architecture, perceived ease of use, flexibility of deployment, quality of documentation, and support accessibility are important factors that can guide developers during selection of a WPS framework that fits a particular application domain best.

Several reviews have been reported in the literature that evaluates spatial services from both a quantitative and qualitative perspectives. MapServer's WMS has been assessed and optimized by Kalbere (2010). COSMC (Czech Office for Surveying, Mapping and Cadastre) and CENIA (Czech Environmental Information Agency) WMSs have been tested for availability and performance (Horák, Ardielli, & Horáková, 2009). Bermudez et al. (2009) compared the ability of WFS and SOS (Sensor Observation Service) to publish time series data. Tamayo et al. (2011) presented an empirical study of instances of servers implementing SOS in terms of compliance with OGC's SWE (Sensor Web Enablement) and interoperability, and in our previous work we evaluated performance of three SOS servers – 52° North, MapServer, and Deegree – based on different test scenarios (Poorazizi, Liang, & Hunter, 2012). Moreover, a WMS performance shootout has been presented annually since 2007 at the FOSS4G (Free and Open Source Software for Geospatial) conference, which provides a standardized procedure for measuring and comparing the performance of WMS server installations (`http://wiki.osgeo.org/wiki/FOSS4G_Benchmark`).

Within the geoprocessing domain, there have been few attempts to evaluate WPS servers. Scholten et al. (2006) investigated efficiency of web services for geoprocessing in a Spatial Data Infrastructure (SDI), but focused on caching, network adaptation, data granularity, and communication modes. Michaelis and Ames (2009) evaluated the WPS 0.4.0 specification, identified challenges, and proposed potential enhancements from an implementation perspective. In addition, a WPS shootout was presented at the FOSS4G conference 2011, which evaluated five WPS servers, 52° North, Deegree, GeoServer, PyWPS, and Zoo, in terms of compliance with OGC's WPS, and interoperability (`http://wiki.osgeo.org/wiki/WPS_Shootout`). The main achievement of the aforementioned works is that they concentrated on influential performance issues, the WPS protocol and its specification, and compliance and interoperability testing. However, there is also a need to evaluate WPS functionality and performance. Through performance evaluation, WPS developers can (i) identify the strengths and weaknesses of each system, and (ii) improve WPS servers to meet both application user and developer requirements (Zhu, 2003). These issues are addressed in this paper. We have evaluated the performance of five WPS servers – 52° North, Deegree, GeoServer, PyWPS, and Zoo – using two test plans in an accessibility assessment scenario. To do so, the WalkYourPlace Transit Model (Steiniger et al., 2013) was used to design the geoprocessing workflow. The workflow was then developed using Python and wrapped and exposed as a standard WPS using the candidate WPS servers. The sample locations were selected using a stratified random sampling approach within the bounds of the City of Calgary, Alberta, Canada. During experiments we controlled the number of concurrent requests, and the WPS input parameters to assess the performance and load capacity of the WPS servers. The remainder of the paper is structured as follows. Section two introduces the WPS specification. The specification of candidate WPS servers is described in section three. Section four explains the methodology used to evaluate the WPS servers, along with a description of the case study, technical architecture, test scenarios, and hardware configuration of the servers used. Section five presents the result. In section six, the WPS servers are assessed in terms of qualitative metrics. Section seven summarizes our findings.

2 Web Processing Service

The OGC released version 1.0.0 of the WPS specification in June 2007 (Schut, 2007). The specification, along with the OGC Web Processing Service Best Practice discussion paper, describe a web service interface that defines how a client and server should cooperate during the execution of a spatial analysis, and how results of the process should be presented (Schäffer, 2012). Clients can send requests via three core operations using three methods: Key Value Pairs (KVP) encoding via HTTP's (HyperText Transfer Protocol) GET, XML (eXtensible Markup Language) via HTTP's POST, or a SOAP/WSDL (Simple Object Access Protocol/Web Service Description Language) approach. The WPS specification defines three mandatory operations that enable spatial processing on the Internet (Schut, 2007). The GetCapabilities operation allows a client to request and receive service metadata documents that describe the capabilities of a specific server implementation. The DescribeProcess operation returns detailed information about a process' requirements, such as input and output parameters, as well as allowable data formats. The Execute operation invokes a specific process implemented by the WPS, using the input parameters provided, and returns the results of the service to a client.

3 WPS Servers

In this paper five WPS servers were used for performance evaluation. 52° North WPS (http://52north.org/communities/geoprocessing/wps/) is developed by the 52° North Initiative for Geospatial Open Source Software GmbH. It implements the three mandatory operations of the WPS 1.0.0 specification. The 52° North WPS server is realized as a servlet and can be deployed in any servlet container such as Apache Tomcat (http://tomcat.apache.org/). Developing a custom WPS process is implemented using 52° North's WPS SDK (Software Development Kit) to define parameters necessary for service configuration, service metadata, and business logic. Spatial analysis functions can be integrated using various libraries such as JTS (http://www.vividsolutions.com/jts/JTSHome.htm), GeoTools (http://www.geotools.org/), R (http://www.r-project.org/), GRASS (http://grass.osgeo.org/), SEXTANTE (http://www.sextantegis.com/), and ArcGIS Server (http://www.esri.com/software/arcgis/arcgisserver), for example.

Deegree WPS (http://www.deegree.org/) is a service built into the Deegree Java framework for geospatial applications and OGC service implementations, deegree 3. deegree 3 is an Open Source Geospatial (OSGeo) Foundation project. It supports the core profile operations of the WPS 1.0.0 standard specification. The Deegree WPS server is implemented as a servlet and can be deployed in any servlet container, i.e., Apache Tomcat. Developing a custom process requires the creation of a Maven (http://maven.apache.org/) project. Configuration parameters and service metadata are defined through XML configuration files and business logic is implemented as a Java class. Deegree WPS currently supports the SEXTANTE spatial library, but other spatial libraries such as FME (http://www.safe.com/fme/fme-technology/) and GRASS (http://grass.osgeo.org/) are being considered.

GeoServer WPS (http://docs.geoserver.org/wps) is part of the popular open-source GIS project GeoServer, a project of the OSGeo Foundation. It supports the three mandatory operations contained in the WPS 1.0.0 specification. The GeoServer WPS server is built using Java technology as a servlet, and runs in an integrated Jetty or Apache Tomcat web server environment. Developing a custom process is accomplished by creating a Maven (https://maven.apache.org/) project. Configuration parameters and service metadata are defined through XML configuration files, and business logic is implemented as a Java class. GeoServer WPS supports GeoTools and JTS spatial libraries.

PyWPS (http://pywps.wald.intevation.org/) is a Python-based WPS implementation developed by Intevation GmbH. It implements the mandatory operations of the WPS 1.0.0 specification. It runs as a CGI (Common Gateway Interface) application and can therefore be deployed in any HTTP Server environment, Apache HTTP Server, for example. Developing a custom process requires the creation a python file to implement the business logic and define service metadata and configuration parameters. PyWPS enables access to a wide range of analysis functions via GRASS, GDAL (http://www.gdal.org/), and R libraries.

Zoo (http://www.zoo-project.org/ is an OSGeo Foundation project that enables existing open source libraries to interact through its WPS framework. It supports the mandatory operations of the WPS 1.0.0 specification. It runs as a CGI application and so can be deployed in any HTTP Server environment. Developing a custom process requires the creation of a configuration file (.zcfg) that defines service metadata and configuration parameters. Business logic can be implemented in several programming languages including C/C++, PHP, JavaScript, Java, Perl, Python, or FORTRAN. Several spatial libraries such as GRASS, GEOS (http://trac.osgeo.org/geos/), and GDAL are supported by default in Zoo WPS framework.

Table 1 lists the technical characteristics of 52° North, Deegree, GeoServer, PyWPS, and Zoo WPS servers.

4 Methodology

In this section, we explain the methodology used to test and measure the performance of the WPS servers.

4.1 Case Study

In order to evaluate performance of the WPS servers, we used the WalkYourPlace Transit Model (http://webmapping.ucalgary.ca/WPSClient/), which is one of the accessibility assessment models developed for the PlanYourPlace project (Steiniger et al., 2013). Based on this model, if the users provide (i) their current location, or perhaps a location they would like to start walking from, (ii) a maximum time they are

	52°North	Deegree	GeoServer	PyWPS	Zoo
Development Platform	Java	Java	Java	Python	C/C++
License	GNU GPL v2	LPGL	GNU GPL v2	GNU GPL v2	MIT/X-11 style
Supported Libraries	JTS GeoTools SEXTANTE R GRASS ArcGIS	SEXTANTE	JTS GeoTools	GRASS GDAL R	GRASS GEOS GDAL
Natively Supported Languages for Process Development	Java	Java	Java	Python	C/C++ Fortran Java Python PHP Perl JavaScript
Service	Servlet	Servlet	Servlet	CGI	CGI
DCP Request	GET, POST, SOAP	GET, POST, SOAP	GET, POST	GET, POST, SOAP	GET, POST

Table 1: WPS servers' technical specifications.

willing to walk to a point of interest, or a transit stop, (iii) average walk-speed, (iv) a maximum time they would like to wait for transit, and (v) and the maximum time they would like to travel by transit, then the system will evaluate the extent of the area that is accessible using pedestrian and transit infrastructure. The services within an accessibility area are then analysed (e.g., point of interests (POI) such as parks, stores, libraries, etc.) to determine an accessibility score for the accessibility area. Should the user wish, they can ask for a distance decay function to be applied that discounts the contribution of POIs that are further away from the users start location. Next, an assessment of crime is undertaken for the accessibility area. The accessibility area, accessibility score, and the crime index are final outputs of the model. For more details about accessibility assessment models deployed as part of the WalkYourPlace framework see Steiniger et al. (2013).

4.2 Technical Architecture

Figure 1 illustrates the processing service architecture for the WalkYourPlace Transit Model. The service architecture has been designed to reduce complexity and enable reuse of geoprocessing services. From a service design perspective, a bottom-up (Granell, Díaz, & Gould, 2010) approach was used to design the services. The geoprocessing services were then implemented using Python in such a way that to be accessible via HTTP GET/POST. In this context, PostGIS spatial functions were used to perform geometric computations such as calculating distances between pairs of points, calculating the area of polygons, and merging multiple geometric objects. Remaining functionality was developed using Python libraries. The geoprocessing services were then wrapped and exposed as standard WPSs using 52° North, Deegree, GeoServer, PyWPS, and Zoo frameworks. In this context, the WPS server acts as a gateway, which enables standard communication with the back-end (Python-based) geoprocessing services. It actually accepts the Execute request, parses the query, and sends it to the corresponding Python-based service using HTTP handlers. After getting the result, the WPS server prepares it as a standard WPS response and sends it back to the client. In this study, we developed seven Python-based geoprocessing modules to perform the analysis, and seven WPS instances using each WPS server to wrap and expose them as standard WPS services (see Figure 1). A PostgreSQL/-PostGIS database was used to store various spatial datasets such as the street and transit networks, the transit schedule, and crime data, obtained originally from OpenStreetMap, Calgary Transit, and the Calgary Police, respectively. To search for attractions within accessibility areas, POI datasets were fetched on demand from OpenStreetMap and MapQuest databases using REST (REpresentational State Transfer) APIs (Application Programming Interfaces). For the calculation of transit-based accessibility areas we used the General Transit Feed Specification (GTFS) formatted data published by the City of Calgary.

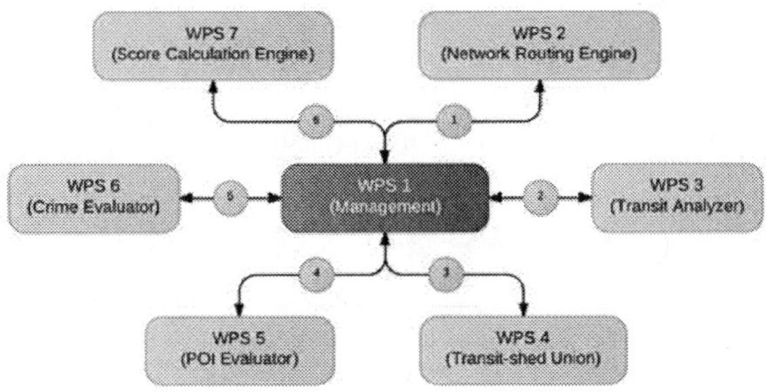

Figure 1. The WalkYourPlace processing service architecture.

The geoprocessing service framework includes an accessibility assessment engine that performs the accessibility analysis through chaining of geoprocessing services in a multi-step pattern, i.e. a workflow. To achieve desired application flexibility, service reusability, and improve performance, the workflow-managed chaining method was used (Alameh, 2003).

Figure 2 presents a UML sequence diagram that outlines how an accessibility score is calculated for pedestrian and transit infrastructure. The client sends a WPS Execute command to the Management WPS, which then initiates an Execute call to the Walkshed WPS. The Walkshed WPS returns a GeoJSON polygon of the network-based accessibility area. The Management WPS then sends an Execute request to the Transit WPS to find all transit stops within the accessibility area and generates an accessibility area for each transit stop based on the user defined constraints described in section 4.1. The Transit WPS returns a GeoJSON-encoded multi-polygon feature. Next, the Management WPS sends an Execute request to the Union WPS to merge all the accessibility areas generated by the Transit WPS. The Union WPS returns a single polygon feature encoded as GeoJSON. The Management WPS then sends an Execute request to the POI WPS to find all attractors within the accessibility area. The POI WPS returns a point set of services encoded as GeoJSON points, along with attributes describing the types of features found. The Management WPS then repeats the same request to the Crime WPS to obtain incident locations. Finally, the Management WPS sends an Execute request to the Aggregation WPS along with the accessibility polygon, the responses from the POI and Crime WPS's, and a Boolean variable to indicate whether the distance decay function should be applied or not. The response from the Aggregation WPS includes an accessibility score, a crime score, and an accessibility area. The Management WPS then returns the Aggregation WPS's response to the client for presentation.

4.3 Test Scenario

In this study, to ensure the same test conditions for all WPS servers were used, we developed the geoprocessing services using Python and then wrapped and exposed them as WPS services. Given this implementation the WPS servers (i.e., 52° North, Deegree, GeoServer, PyWPS, and Zoo) act as a gateway that enables standard interaction between clients (i.e., the user or other services) and back-end geoprocessing services, which implemented using Python. For example, when the client sends an Execute request to the Management WPS, it then sends a request to a corresponding Python service, which is accessible via HTTP GET/POST. After getting the response, the Management WPS sends Execute requests to other WPS services (i.e., Walkshed, Transit, Union, POI, Crime, and Aggregation WPSs), which in turn communicate with back-end Python services to get the processing result. As such, the Execute method depends on external service calls, and the response time for invocation of the whole workflow (Figure 2) was measured to evaluate the "end-to-end" performance, i.e., the response time includes communication time and processing time.

To evaluate the performance of the WPS servers, we designed two test scenarios based on the accessibility assessment case study. In the first scenario (Scenario A), we randomly chose the WPS input parameters to generate 45 Execute requests. The number of concurrent requests was assumed constant (n=1). The input parameters were selected using the following criteria:

- Walking Start Point: sample locations were selected using a stratified random sampling approach within the bounds of the City of Calgary (see Figure 3).
- Walking Start Time: random timestamps between 5 a.m. and 12 p.m., which is the Calgary Transit hours of operation (`http://www.calgarytransit.com/accesscalgary/hours.html`).
- Walking Time Period: we selected random values between 5 minutes and 20 minutes.

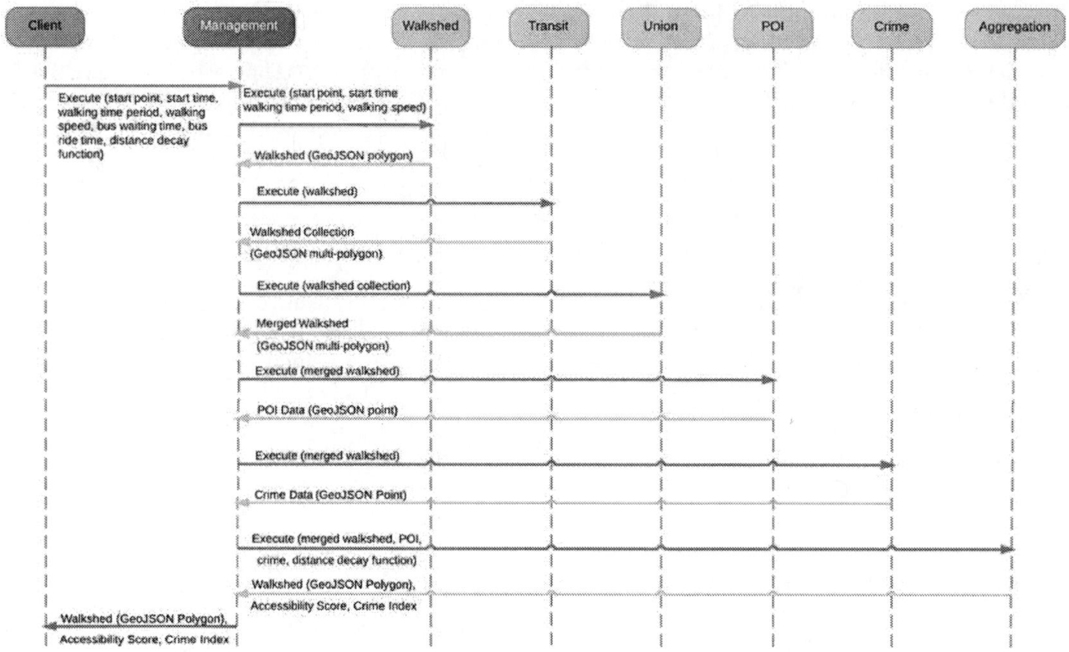

Figure 2. UML activity diagram of accessibility assessment workflow.

- Walking Speed: we selected random values between 3 km/h and 6 km/h with step values of 0.5 km/h.

- Bus Waiting Time: we selected random values between 0 minutes and the Walking Time Period.

- Bus Ride Time: we selected random values between 0 minutes and Walking Time Period – Bus Waiting Time.

- Distance Decay Function: a Boolean variable (i.e., True/False) was selected randomly.

For the second scenario (Scenario B), we focused on the number of concurrent requests. In this context, the number of concurrent requests was generated using a 2^n pattern, while variable "n" was selected between 0 and 7 with step value of 1. 30 WPS Execute requests were generated for each WPS service and replicated according to the concurrent request pattern. All other criteria were determined using the above mentioned approach for Scenario A.

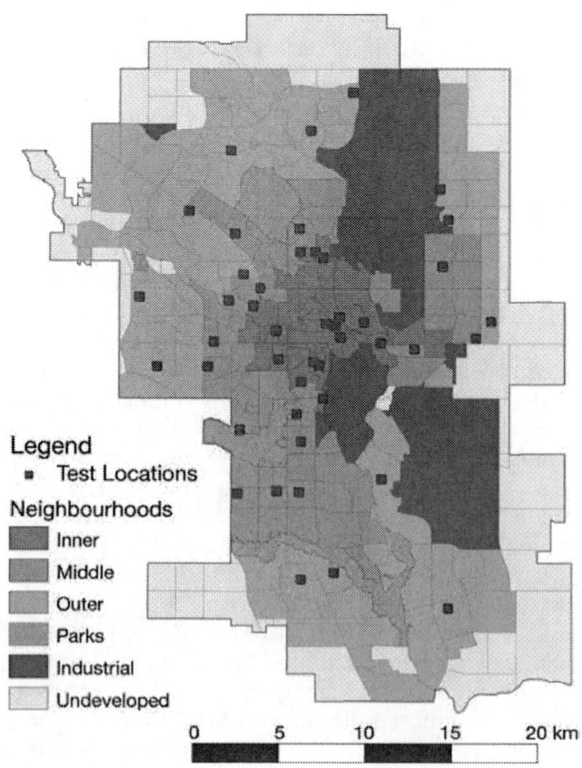

Figure 3. Map of the City of Calgary highlighting the locations used for evaluating the WPS servers.

4.4 Test Environment

To more accurately reflect the users experience, all the tests have been measured from the client-side. On the server-side, a Dell OptiPlex 990 was used as the host machine, with an Intel Core i5 (3.1GHz) CPU, 8GB of RAM, and 500GB of disk space, running Microsoft Windows 7 Professional (64-bit). In order to deploy and test the WPS servers under the same conditions, each WPS package was installed on a separate virtual machine with the same hardware configuration. VMware Player 5.0.1 (`http://www.vmware.com/`) was used to setup five virtual machines with access to 4GB of RAM, 40GB of disk space, and use of 4 out of 8 CPU cores, running Ubuntu 12.04 LTS (64-bit). The network configuration of the virtual machines was set to "Bridged", allowing them to connect directly to the physical network and obtain a dedicated IP address. Table 2 summarizes the configuration of the server machine (host), and virtual machines. For more information about the configuration of database server and software libraries used see Appendix A.

Hardware	**Dell OptiPlex 990 (Host)**	**VMware (VM)**
CPU	Intel Core i5 3.1GHz	4 Cores of 8
RAM	8GB	4GB
HDD	500GB	40GB
OS	Windows 7 Professional (64-bit)	Ubuntu 12.04 LTS (64-bit)

Table 2. Experimental server configuration.

The machine used to run the tests at the client-side was the host machine. In this study, we used the same machine to set up the servers and test them, while according to (VMware, 2006), "an ideal setup for workloads that involve network traffic is to use an external client (on a different physical system) to send network traffic to and receive network traffic from a virtual machine". Although this could affect the performance of the WPS servers, the test conditions (i.e., hardware and software configurations) were the same for all the servers, which are shown in Table 2, Table 6, and Table 7. It was assumed that network time would be constant and therefore would not contribute significantly to differences in response times.

In order to run the tests and measure performance factors (e.g., response time, response size, etc.), Apache JMeter (`http://jmeter.apache.org/`) was used, as it is a widely accepted performance-testing tool for web applications.

5 Experimental Results

Since each WPS server uses database connections to execute queries, a warm-up run was first performed. This ensures that the overhead of establishing a connection to the database is not accounted for in the metrics (elapsed time). During each performance test, only one virtual machine was run. Response time, response size, and whether or not a request was successful were logged. This data allowed the estimation of average response times, average server throughput, average server failure rate, and average response size returned by each WPS server. Figure 4 to Figure 7 and Table 3 below report the results of the experiments.

First, the performance test for Scenario A is reported. The average response time, time taken for a service call to return all response bytes, the average response size, the quantity of data exchanged between client and server, for each of the WPS servers are listed in Table 3 and plotted on Figure 4.

Given the data, the most rapid WPS server was Deegree, with an average response time of 2.499 $\pm$ 1.259 s (95% confidence interval (CI)), followed by GeoServer WPS, 52° North WPS, Zoo WPS, and PyWPS. A one-way Analysis of Variance (ANOVA) test indicates that all WPS servers respond similarly with no significant difference between them, $F(4,220)=0.739$, $p=0.566$.

In terms of response size, there was no significant difference ($F(4,220)=1.071$, $p=0.372$) either with all WPS servers returning similar response package data volumes ($\approx$ 2.484 kB). The GeoServer WPS returned the least amount of data (2.301 $\pm$ 0.267 kB) to the client, and PyWPS had the most (2.686 $\pm$ 0.269 kB).

The reason of having different response sizes was because of a slight different in XML tags within the Execute response. For example, wps:ExecuteResponse content is listed in Table 4 for PyWPS and GeoServer WPSs' Execute response, which returned the most, and the least amount of data, respectively.

Scenario B was designed to assess the effect of increased load on each server. The effect of load was assessed by increasing the number of concurrent requests from 1, to 2, 4, 8, 16, 32, 64, and finishing with 128 concurrent requests. Individual services and the service chain were tested using pre-defined input parameters under normal condition (n=1) and no error was observed. Those parameters were then used to measure the performance of the WPS servers under high loads (n>1). To get representative results, all of the experiments were repeated 30 times and the response time, response size, and server success/failure were recorded. These data allowed the estimation and comparison of server throughput. The results are depicted in Figure 5 to Figure 7.

WPS Server	Response Time (s)	Response Size (kB)
52° North	2.784 ± 1.269	2.448 ± 0.267
Deegree	2.499 ± 1.259	2.505 ± 0.267
GeoServer	2.753 ± 1.255	2.301 ± 0.267
PyWPS	3.995 ± 1.661	2.686 ± 0.269
Zoo	2.999 ± 1.313	2.479 ± 0.269

Table 3. Results for Execute request (Scenario A).

Figure 5 shows that Deegree, GeoServer, and Zoo generally perform similarly, the only difference is an improvement in response time by Deegree for 128 concurrent requests. With an increase from 64 to 128 concurrent requests Deegree's response time improves to approximately half that of PyWPS and Zoo. It is apparent that 52° North and PyWPS had difficulty when more than 16 and 64 concurrent requests were received for processing respectively. It is also evident that when more than 64 concurrent requests were sent to PyWPS and Zoo WPS servers failure rates increased dramatically, approaching 100% at 128 concurrent requests. Throughput was also affected significantly by the number of concurrent requests, particularly for 52° North, which returned less than 1 successful request per hour once concurrent requests increased above 16. All servers performed substantially better when only one request was received at a time, with 52° North achieving a throughput of 1,445 successful requests per hour, followed by Zoo with 1,145, GeoServer with 1,115, Deegree with 1,024, then PyWPS with 894 requests per hour.

Because of the variation in the data, when analyzing the results using a two-way ANOVA, only the number of concurrent requests had an effect on load testing ($F(1,30)=20.640$, $p<0.001$), individual servers did not contribute to differences observed. As the number of concurrent requests increased GeoServer and Zoo followed a similar (linear) trend. Deegree tended to perform better, especially under high loads (n=128). In addition, 52° North and PyWPS failed to respond while processing more than 16 and 64 requests respectively. The failure rate of Deegree and GeoServer exhibited a same pattern. We observed a failure rate of 0.8% under high loads (n > 4). PyWPS and Zoo also followed a same failure rate pattern. It was constant (≈1.6%) between four and 64 concurrent requests and then approached 100% under higher loads (n > 64). All the WPS servers performed similarly in terms of throughput, for example with four concurrent requests they processed around 600 requests per hour. It suggests that the WPS servers were capable of handling a request every six seconds (n = 4). This result requires further investigation to determine if the servers can be tuned to function more effectively under real-world conditions. These results are summarized in Figure 5 to Figure 7.

6 Lessons Learned

In this section, the relative advantages and disadvantages of each WPS server, and challenges experienced while working with them are discussed. In this context, the WPS servers were evaluated from a qualitative perspective in terms of: ease of installation and configuration; perceived ease of use and flexibility for creating new processes; native support for development languages; quality of documentation; and community support. The qualitative comparison results are shown in Table 5.

Installation – as 52° North WPS, Deegree WPS, and GeoServer WPS servers are servlet-based applications, the installation process was straightforward. For 52° North and Deegree, installation is accomplished by deploying the downloaded/built WAR (Web ARchive) file into a servlet container such as Apache Tomcat. For GeoServer, after deploying the WAR file into a servlet container, the WPS Extension should be extracted to the WEB-INF/lib directory of the GeoServer installation. Library dependency was the main issue with PyWPS and Zoo WPS servers' installation process. They have several library dependencies that must be installed first. PyWPS follows a typical Python installation procedure using a setup.py script. Further configuration is necessary to set server paths, and the process folder locations. Installation of the Zoo Kernel, configuration, and installation of the Zoo Service Provider were the main steps required to deploy a service on the Zoo WPS server.

Creating a new process and configuration – as 52° North WPS, GeoServer WPS, and Zoo WPS

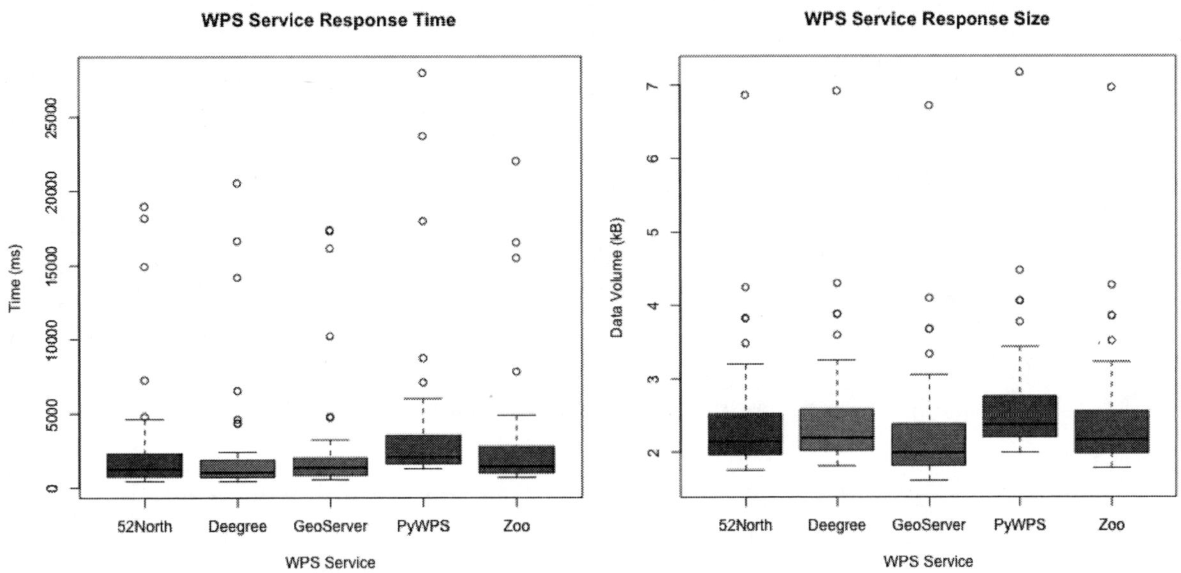

Figure 4. Response time (left) and size (right) for Execute requests (Scenario A).

WPS Server	The Execute Response
PyWPS	<wps:ExecuteResponse xmlns:wps="http://www.opengis.net/wps/1.0.0" xmlns:ows="http://www.opengis.net/ows/1.1" xmlns:xlink="http://www.w3.org/1999/xlink" xmlns:xsi="http://www.w3.org/2001/XMLSchema-instance" xsi:schemaLocation="http://www.opengis.net/wps/1.0.0 http://schemas.opengis.net/wps/1.0.0/wpsExecute_response.xsd" service="WPS" version="1.0.0" xml:lang="en-CA" serviceInstance="http://localhost/cgi-bin/wps?service=WPS&request=GetCapabilities&version=1.0.0" statusLocation="http://localhost/wpsoutputs/pywps-137824772530.xml">
GeoServer	<wps:ExecuteResponse xml:lang="en" service="WPS" serviceInstance="http://10.146.29.24:8080/geoserver/ows?" version="1.0.0" xmlns:wps="http://www.opengis.net/wps/1.0.0" xmlns:ows="http://www.opengis.net/ows/1.1" xmlns:xlink="http://www.w3.org/1999/xlink">

Table 4: A portion of the Execute response document returned by PyWPS and GeoServer WPS.

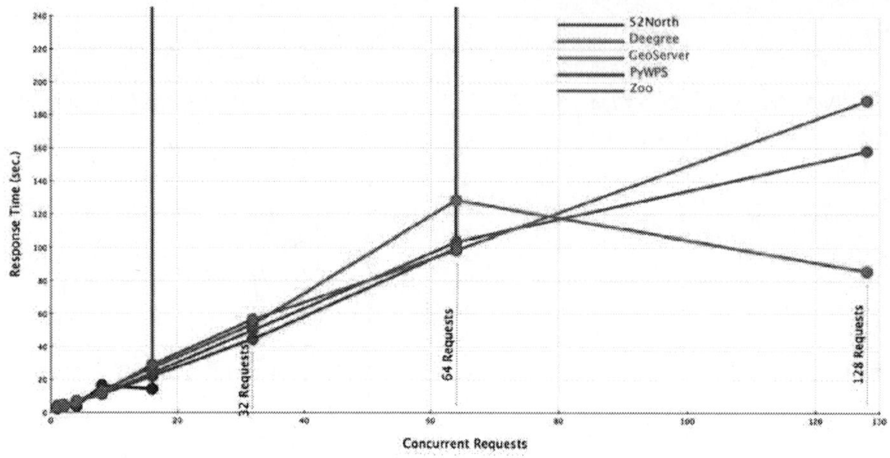

Figure 5. Response time when increasing concurrent requests.

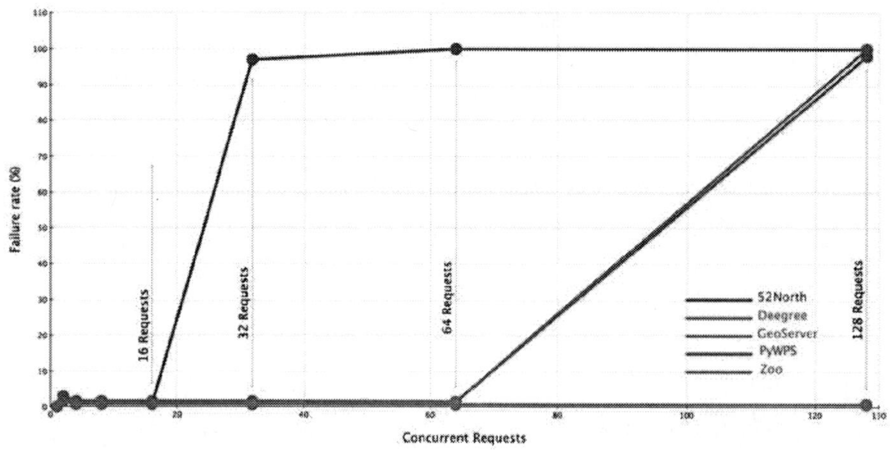

Figure 6. Failure rate with increasing concurrent requests.

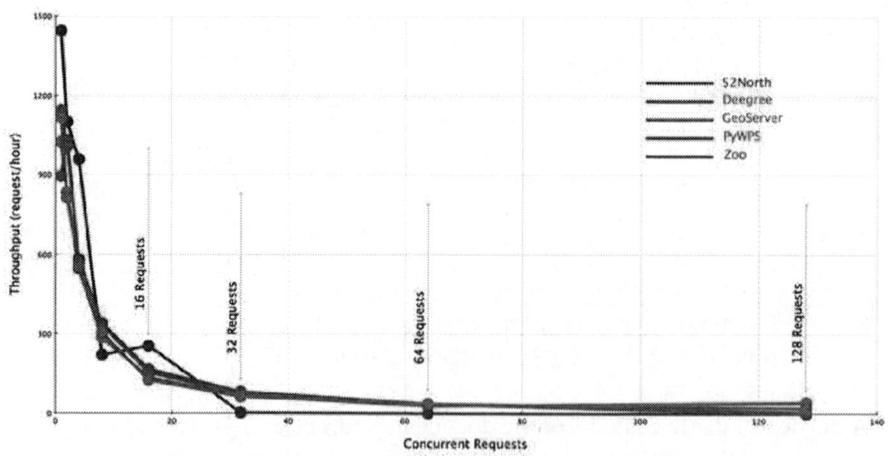

Figure 7. Throughput with increasing concurrent request.

	52°North	Deegree	GeoServer	PyWPS	Zoo
Installation*	Easy	Easy	Easy	Difficult	Difficult
Create new processes and Configuration*	Easy	Medium	Easy	Difficult	Easy
Native Development Languages	Java	Java	Java	Python	C/C++ Fortran Java Python PHP Perl JavaScript
Quality of Documentation**	Great	Good	Great	Good	Good
Community Support	Mailing list, Wiki, Forum, Issue Tracker, SVN, GitHub	Mailing list, Wiki, Forum, Issue Tracker, SVN, GitHub	Mailing list, Forum, Issue Tracker, SVN, IRC Meeting, GitHub	Mailing list, GitHub	Mailing list, Forum, Issue Tracker, SVN, GitHub

* Ranking ranges: Easy; Medium; Difficult

** Ranking ranges: Weak; Good; Great

Table 5. WPS servers and their features from a qualitative perspective.

frameworks were well-documented, a new process was simple to create and easy to configure. For 52° North WPS, this procedure was accomplished using the WPS SDK in three steps: (i) create a Java class for the process, (ii) export the process as a JAR (Java ARchive) file, and (iii) deploy the process into 52° North's WPS framework. For GeoServer WPS, a new process is developed by creating a Maven project in three steps: (i) create a Java class and an XML configuration file for the process, (ii) compiling the project as a JAR file, and (iii) deploying the process into GeoServer's WPS framework. To create a new process for Zoo WPS, two steps have to be completed: (i) create a service file using one of the supported programming languages, and a zcfg configuration file for the process, and (ii) deploy the CGI application into Zoo's WPS framework. Although Deegree's documentation (`http://download.deegree.org/documentation/3.3.3/html/`), was well-organized and comprehensive, it was not particularly clear how to build and deploy a new process within Deegree's WPS framework, nor were there many examples to base development on. However, a Maven project should be created and three steps should be followed to add a new process to Deegree's WPS: (i) create a Java class and an XML configuration file for the process, (ii) compile the project as a WAR file, and (iii) deploy the servlet application into any servlet container. To add a new process to PyWPS framework, two steps should be followed: (i) create a service file and modify the configuration files (i.e., pywps.cfg and pywps.cgi), and (ii) deploy the CGI application into PyWPS's framework. On occasion the PyWPS server returned an HTTP Error 500 that prevented it from fulfilling WPS requests, especially after a new process had been added. To resolve this failure several access permission settings were required (for more details see Hamre (2011)).

Native Development languages – 52° North WPS, Deegree WPS, GeoServer WPS, and PyWPS frameworks support one native programming language each for the development of a new process, while the Zoo WPS framework supports seven programming languages. This adds flexibility for developers, as they are able to either develop new processing services in their language of choice, or develop services as independent modules that may draw on libraries from many different languages.

Quality of documentation – 52° North WPS and GeoServer WPS documentation was comprehensive and provide clear instruction for installation and configuration of the WPS servers, along with clear instructions for developing new process instances.

Community support – 52° North WPS, Deegree WPS, GeoServer WPS, and Zoo WPS frameworks have large communities of users/developers and provide different communication mediums to support them. PyWPS does not appear to have an active community of users/developers, which may make access to support difficult.

7 Discussion and Conclusions

We have evaluated performance of WPS servers using two test scenarios via a case study that focuses on accessibility assessment. In the first scenario, the WPS servers were tested using 45 randomly generated Execute requests, holding the number of concurrent requests constant (n=1). The results show that on average Deegree returns the response package most rapidly. However, a one-way ANOVA test showed that, given the data, there is no significant difference in response time between the WPS servers tested ($F(4,220)=0.739$, $p=0.566$), nor data volume returned ($F(4,220)=1.071$, $p=0.372$).

In the second scenario, load testing was undertaken by varying the number of concurrent requests. Overall Deegree and GeoServer performed similarly, although Deegree tended to perform better under high loads. 52° North had difficulty when more than 16 concurrent requests were received for processing, but performed more effectively under low loads compared to other WPS servers. Under low loads, n=1, 52° North had the highest throughput completing 1,445 requests per hour, followed by Zoo with 1,145 requests per hour, GeoServer with 1,115 requests per hour, and Deegree and PyWPS completing 1,024 and 894 requests per hour respectively. Throughput for 52° North effectively went to zero requests per hour once the load increased to more than 16 concurrent requests. Although no failed requests were encountered under low loads, n=1, success rate for Deegree and GeoServer stabilized at four or more concurrent requests, to approximately 99.2%. PyWPS and Zoo followed the same pattern, with a success rate of 98.4% between four and 64 concurrent requests.

While four CPU cores were allocated to each WPS server during testing, upon reviewing CPU load logs it was evident that, except for PyWPS, only one CPU core was generally being used at any time during testing. Specifically, Deegree used only one CPU core; 52° North, GeoServer, and Zoo each used around 20% of one CPU core and 5% of the other cores. PyWPS used all cores during testing. In addition, memory usage of all WPS servers was constant (with minor fluctuations) during testing. On average memory use was 30%. This suggests that performance improvements may be possible if server specific tuning, or more effective development strategies are implemented. For example, the use of multiple CPU cores in Java-based applications is handled via JVM (Java Virtual Machine), which generally tends to be problematic. In this context, if particular implementation approaches or software libraries (e.g., concurrency libraries) are used it may result in a better performance.

We must also note that a WPS server's response time is dependent upon the intensity of a service's processing requirements. As such, performance results will depend on the complexity of the workflow, the complexity of individual back-end processes, and the complexity of the data.

The WPS servers have also been assessed in terms of qualitative metrics. 52° North WPS, Deegree WPS, and GeoServer WPS servers are easy-to-install and are well documented. They also have worldwide communities of developers/users, and provide different ways of communication to support their users/developers. The documentation for PyWPS was not complete, nor was it always clear and concise, making it difficult to install and configure the PyWPS server. PyWPS does not appear to have an active community of users/developers, and users/developers, as a consequence, may suffer from lack of support. Zoo WPS does have accessible documentation and an accessible support community. It also supports several programming languages and offers powerful and flexible approaches to develop WPS instances. Generally, compared to other WPS servers, 52° North and GeoServer seem to be the best choices when considering qualitative metrics, as they met most of the evaluation criteria we chose in this study.

It should be noted that standard compliance is a major issue in the WPS domain, which was not investigated in this study. Interoperability and standard compliance tests can be undertaken as a part of qualitative evaluation process, which focuses on schema, semantics, and encodings.

To conclude, when selecting an appropriate WPS server, we believe it is important to consider both quantitative and qualitative metrics. The importance of each metric can be weighted based on different application requirements. Generally speaking, from a user's perspective, performance is one of the most important factors when choosing a web-based application, while from developers' perspective, qualitative factors such as perceived ease of installation and configuration, variety of development languages, quality of documentation and accessibility of support may be more critical. To choose a WPS server, we suggest starting an evaluation process with a basic set of questions that are linked to the evaluation criteria. The questions could be "who is the user of the system?" "What should the end-user be able to do with the system?" "What programming languages are developers comfortable with for develop of the system?" "How complex are the back-end processes?" "How should the system function, synchronous or asynchronous?" "What is the architecture used to design the processing workflow?" "What is the expected number of users?" In the end, the most appropriate WPS server should be selected based on a trade-off between quantitative performance metrics and qualitative "ease of use" metrics for a specific application or use case. This may lead to the selection of different WPS servers for different applications.

Note: the developed Python-based geoprocessing services, WPS instances, and test scripts are publicly available, see Appendix B.

References

Alameh, N. (2003). Chaining geographic information Web services. IEEE Internet Computing, 7(5), 22 - 29. doi:10.1109/MIC.2003.1232514

Bermudez, L., Cook, T., Forrest, D., Bogden, P., Galvarino, C., Bridger, E., ... Graybeal, J. (2009). Web feature service (WFS) and sensor observation service (SOS) comparison to publish time series data. In Proceedings of International Symposium on Collaborative Technologies and Systems, 2009 (CTS '09) (pp. 36 –43). Baltimore, MD , USA. doi:10.1109/CTS.2009.5067460

Cibulka, D. (2013). Performance Testing of Web Map Services in three Dimensions – X, Y, Scale. Slovak Journal of Civil Engineering, XXI(1). doi:10.2478/sjce-2013-0005

De La Beaujardiere, J. (2004). Web Map Service (No. OGC 04-024). Open Geospatial Consortium Inc. Retrieved from portal.opengeospatial.org/files/?artifact_id=5316

Dubois, G., Schulz, M., Skøien, J., Bastin, L., & Peedell, S. (2013). eHabitat, a multi-purpose Web Processing Service for ecological modeling. Environmental Modelling & Software, 41, 123–133. doi:10.1016/j.envsoft.2012.11.005

Evans, J. D. (2003). Web Coverage Service (WCS), Version 1.0.0 (No. OGC 03-065r6). Open Geospatial Consortium Inc. Retrieved from portal.opengeospatial.org/files/?artifact_id=3837

Granell, C., Díaz, L., & Gould, M. (2010). Service-oriented applications for environmental models: Reusable geospatial services. Environmental Modelling & Software, 25(2), 182–198. doi:10.1016/j.envsoft.2009.08.005

Hamre, T. (2011, December 29). Open Service Network for Marine Environmental Data - WPS Cookbook. http://netmar.nersc.no/. Retrieved from http://netmar.nersc.no/sites/netmar.nersc.no/files/NETMAR_D7.7_WPS_Cookbook_r1_20111229.pdf

Horák, J., Ardielli, J., & Horáková, B. (2009). Testing of web map services. International Journal of Spatial Data Infrastructures Research,(Special Issue: GSDI 11), 25. Retrieved from http://www.gsdi.org/gsdiconf/gsdi11/papers/pdf/330.pdf

Kalbere, P. (2010). Performance and statistical analysis of WMS servers. In Proceedings of FOSS4G 2010. Barcelona, Spain.

Mayer, C., Stollberg, B., & Zipf, A. (2009). Providing Near Real-Time Traffic Information within Spatial Data Infrastructures. In Proceedings of International Conference on Advanced Geographic Information Systems Web Services (GEOWS '09) (pp. 104 –111). Cancun, Mexico. doi:10.1109/GEOWS.2009.17

Michaelis, C., & Ames, D. (2009). Evaluation and Implementation of the OGC Web Processing Service for Use in Client-Side GIS. GeoInformatica, 13(1), 109–120. doi:10.1007/s10707-008-0048-1

Nebert, D., Whiteside, A., & Vretanos, P. (2007). OpenGISő Catalogue Services Specification (No. OGC 07-006r1). Open Geospatial Consortium Inc. Retrieved from portal.opengeospatial.org/files/?artifact_id=20555

Poorazizi, M. E., Liang, S. H. L., & Hunter, A. J. S. (2012). Testing of sensor observation services: a performance evaluation. In Proceedings of the First ACM SIGSPATIAL Workshop on Sensor Web Enablement (pp. 32–38). Redondo Beach, CA, USA.: ACM. doi:10.1145/2451716.2451721

Schäffer, B. (2012). Web Processing Service Best Practices Discussion Paper. Open Geospatial Consortium.

Scholten, M., Klamma, R., & Kiehle, C. (2006). Evaluating Performance in Spatial Data Infrastructures for Geoprocessing. IEEE Internet Computing, 10(5), 34–41. doi:10.1109/MIC.2006.97

Schut, P. (2007). OpenGIS Web Processing Service (No. OGC 05-007r7). Open Geospatial Consortium Inc. Retrieved from portal.opengeospatial.org/files/?artifact_id=28772&version=2

Solntseff, N., & Yezerski, A. (1974). A survey of extensible programming languages. Annual Review in Automatic Programming, 7, Part 5, 267–307. doi:10.1016/0066-4138(74)90001-9

Steiniger, S., Poorazizi, M. E., & Hunter, A. J. S. (2013). WalkYourPlace - evaluating neighbourhood accessibility at street level. In Proceedings of the 29th Urban Data Management Symposium (Vol. XL–4/W1). London, UK: ISPRS International Archives of Photogrammetry, Remote Sensing, and Spatial Information Science.

Tamayo, A., Viciano, P., Granell, C., & Huerta, J. (2011). Empirical Study of Sensor Observation Services Server Instances. In S. Geertman, W. Reinhardt, F. Toppen, W. Cartwright, G. Gartner, L. Meng, & M. P. Peterson (Eds.), Advancing Geoinformation Science for a Changing World (Vol. 1, pp. 185–209). Springer Berlin Heidelberg.

VMware. (2006). Performance Benchmarking Guidelines for VMware Workstation 5.5. VMware, Inc.

Vretanos, P. A. (2002). OpenGIS Web Feature Service Implementation Specification. Open Geospatial Consortium Inc.

Zhu, J. (2003). Quantitative Models for Performance Evaluation and Benchmarking: Data Envelopment Analysis With Spreadsheets and Dea Excel Solver. Springer.

Appendix A

Hardware	Dell OptiPlex 960
CPU	Intel Core 2 Quad 3.0GHz
RAM	8GB
HDD	500GB
OS	Ubuntu 13.04 (64-bit)

Table 6. Experimental database server configuration

Software	**Version**
52° North	3.2.0
Deegree	3.0.4
GeoServer	2.4.3
PyWPS	3.2.1
Zoo	1.3.0
Java	Oracle JDK 7
Servlet Container	Apache Tomcat 7.0.30
Python	2.7.3
PostgreSQL/PostGIS	9.1.12/1.5.3

Table 7. Software libraries used to setup WPS servers

Appendix B

The developed Python-based geoprocessing services, WPS instances, and test scripts are publicly available at the following URLs:

Test Scripts:

- https://github.com/mepa1363/foss4g-test-script

Python-based geoprocessing services:

- https://github.com/mepa1363/wyp-server-52north-foss4g
- https://github.com/mepa1363/wyp-server-deegree-foss4g
- https://github.com/mepa1363/wyp-server-geoserver-foss4g
- https://github.com/mepa1363/wyp-server-pywps-foss4g
- https://github.com/mepa1363/wyp-server-zoo-foss4g

WPS instance:

- https://github.com/mepa1363/wyp-wrapper-52north-centralized-transit
- https://github.com/mepa1363/wyp-wrapper-deegree-centralized-transit
- https://github.com/mepa1363/wyp-wrapper-geoserver-centralized-transit
- https://github.com/mepa1363/wyp-wrapper-pywps-centralized-transit
- https://github.com/mepa1363/wyp-wrapper-zoo-centralized-transit

GRASS GIS, Star Trek and old Video Tape

by Peter Heinz Löwe[1]*, Janna Neumann*[1]*, Margret Plank*[1]*, Frauke Ziedorn*[1]*, Robert Lozar*[2]*, James Westervelt*[2] *and Roger Inman*[3]

[1]*National Library of Science and Technology (Germany),* [2]*ERDC-RDE-CERL (USA),* [3]*Movingpictures TV (USA).* peter.loewe@tib.uni-hannover.de

Abstract

This paper discusses the need for the preservation of audiovisual content in the OSGeo communities beyond the established software repositories. Audiovisual content related to OSGeo projects such as training videos can be preserved by multimedia archiving and retrieval services which are currently developed by the library community. This is demonstrated by the reference case of a newly discovered version of the GRASS GIS 1987 promotional video which is being included into the AV-portal of the German National Library of Science and Technology (TIB). Access to the video will be provided upon the release of the web-based portal, allowing for extended search capabilities based on enhanced metadata derived by automated video analysis. This is a reference case for future preservation activities regarding semantic-enhanced Web2.0 content from OSGeo projects.

Keywords: GRASS GIS, OSGeo, digital preservation, educational material, audio visual media, Youtube, GRASS 1987 promotional video, Digital Object Identifiers, audiovisual history, screen casts, Web 2.0, Multimedia retrieval.

1 Knowledge Preservation in the OSGeo Communities

1.1 The Role of OSGeo

Since its launch in 2006, the Open Source Geospatial Foundation (OSGeo) has distinguished itself as an umbrella organisation, incubation tank and software license clearinghouse for a large and growing number of geospatial Free and Open Source (FOSS) software projects (OSGeo 2014).

Work in these projects is done by international communities of volunteers. It is centered, but not limited to the development of software tools. Tasks like software testing, the creation of reference data, technical writing for user-and developer-manuals, multi-language translation and the creation of tutorials/educational material augment the core software development activities. Without these tasks, the access to the software would be seriously hampered for the majority of users.

1.2 Use of Repositories and Web 2.0 use in OSGeo projects

While the software of the OSGeo projects is maintained in repository systems such as CVS, SVN and Git, most of the audiovisual educational material is currently provided via virtual Web 2.0 communities, including Slideshare and YouTube. References to the content are made by links and free classification (Folksonomy/Tagging). For the licensing of this intellectual property are often Creative Commons licenses used.

The content which is shared on the Web2.0 channels consists of experience gained with specific software instances for geospatial analysis or processing tasks. This is an important source of practical know-how for Geo–informatics practitioners.

1.3 The Challenge of Audiovisual Content Preservation

The audiovisual content provided through the Web 2.0 channels continues to grow for all OSGeo projects. With the ubiquity of screen capture software and video recording, this approach has distinguished itself as a fast and affordable alternative to preserve the underlying knowledge in text documents. Collaborative tagging is used to provide searchable keywords regarding the actual content. While this is sufficient to search for the names of specific OSGeo software projects, it is an ineffective means to query specific software versions or the description of complex or specialized workflows. Until now, there are no explicit community rules or best practices how long such geospatial-themed audiovisual content will be kept available. It may be eventually removed by its creator without previous notice, but might also go offline once the Web 2.0 portal is retired.

The discussion of long term preservation of audiovisual content and effective search access for audiovisual content within the scope of OSGeo has just begun. Most content providers still consider the Web 2.0 portals as ubiquitous untrustworthy providers of persistent storage space. However, this assumption remains to be verified.

Both from the perspective of the OSGeo communities and research libraries, it is imperative that the knowledge and scientific expertise provided through this audiovisual content is preserved, made fully searchable and citable for future reference. For this reference cases and best practices are needed.

Year	GRASS Installations (Sites)	GRASS Version
1982	1	-
1983	3	-
1984	5	-
1985	20	GRASS 1.0
1987	100+	GRASS 3.0
1988	1000+	GRASS 3.0
1989	1000+	GRASS3.1 (public domain)
1991	1000+	GRASS 4.0 (ftp: 128.174.5.50)

Table 1: Growth of GRASS installations from 1982 – 1991 (Westervelt 1991).

In 1989, the software was placed in the Public Domain and was made available on the Internet via anonymous FTP starting in 1991. The licensing under the General Public License (GPL), beginning with GRASS5.0 in 1999 has resulted in a strong growth of the developer community, leading to new features which grew the user base further worldwide. The license model remains unchanged for the current GRASS6.0 versions and the upcoming GRASS7.0.

During its long development, GRASS has attracted multiple generations of users and developers. While development began in 1982 on Z-80 8-bit CPUs with 64 KB address space, the software was soon ported to improved hardware platforms and operating systems like UNIX. Since 2005, 64bit CPUs are natively supported. Currently GRASS GIS is available for a wide range of computing environments, spanning from Android-based palmtops over desktop PCs to High Performance Computing Clusters (Neteler 2013, Löwe et al. 2012).

A side effect of the GRASS development effort was creation of the modern Open Geospatial Consortium (OGC). It was originally founded in 1987 under the name Open GRASS Foundation (OGF), taking the project lead from the U.S. Army Corps of Engineers Construction Engineering Research Laboratory (USA CERL) (Westervelt 2004).

2 Geographic Resource Analysis Support Software (GRASS) GIS

2.1 GRASS GIS Development Overview

GRASS GIS, the Geographic Resource Analysis Support Software, is one of the oldest Free and Open Source Geographic Information System (GIS) projects (GRASS GIS 2014). The acronym was introduced to adhere to the common use of plant names in earlier Geographic Information Systems (GIS), such as SAGE and MOSS (Mapping Overlay Statistical System) (Westervelt 2004). While being a founding project of OSGeo, it predates the organization by several decades, having been under continuous development since 1982.

From its launch in 1982 until 1997, GRASS GIS was hosted at U.S. Army, Corps of Engineers Research Laboratory (USA CERL). Baylor University, Texas, maintained the software from 1997 to 1999. Beginning in 1998 GRASS GIS was hosted at University of Hannover, Germany until 2001, when ITC-irst in Trento, Italy took over. In 2006, GRASS GIS became one of the first projects to join the OSGeo Foundation. Since then its main repository is hosted by the OSGeo in the USA.

Since the beginning of the project, the GRASS user base has continuously grown. From 1982 to 1991 the dropping prices for computer equipment were the driving factor. In this decade before the advent of the WWW, the user base of GRASS was measured in "sites" installations. See table 1 for details.

2.2 "GRASS -the adventure begins" – the 1987 Video Commercial

In 1987 a video commercial was produced by the U.S. Army Natural Resources Management Program to promote the use of GRASS GIS. By that time, the user base had grown to over hundred installation sites, using GRASS 2.0 on hardware which required an investment of 40,000 USD.

2.2.1 Filming and Production

Production took six months and was managed by Robert Lozar as the Principal Investigator. Filming and special effects were carried out by Moving Pictures Productions Champaign Illinois. For audio, a soundtrack was composed by Scott Wyatt and the audio script was narrated by the professional actor William Shatner.

2.2.2 Legacy

Following its commercial use, the promotional video remained unavailable to the growing worldwide GRASS GIS community until 2004. A digitized copy of a remaining analog VHS tape was shown at the FOSS/GRASS Users Conference 2004 in Bangkok. Subsequently, this digitized version was made available for download from the GRASS project website and is currently available from the OSGeo portal (OSGeo: GRASS MOVIE CERL 2014) and several repostings on YouTube (Youtube: GRASS MOVIE CERL 2014).

2.2.3 Content and Significance

The content of the promotional video provides a generic introduction to the basic concepts and potential applications of geographic information systems to land managers in the 1980s, emphasizing the benefits of the use of GRASS GIS. In addition, projections for decreasing hardware costs for GIS installations are given and supported hardware platforms are listed.

The video is a rare piece of documentation from the early days of GIS and Geoinformatics. As the development of GRASS GIS still continues 27 years after the production of the video, it is noteworthy that for many command sequences shown in the video modern counterparts still exist. Also, the video indicates that the Spearfish sample data set (Spearfish Sample Data 2014) still provided for

GRASS GIS can be traced back to 1987. Since its re-release in 2004, the promotional video is used both in GIS education and OSGeo events to emphasize the rapid growth of computer processing power and storage space.

2.2.4 The Citation Problem

Until now, no permanent way to reference the promotional GRASS video and cite its content exists. The WIRED internet magazine addressed this issue in an article explicitly in 2013, stating that the video is not referenced on the International Movie Database or Wikipedia. It is noteworthy that the article provides a YouTube-based link to the video, but not to the main site at the OSGeo portal (Mason 2013).

2.2.5 Star Trek and GRASS GIS

The TV show Star Trek (IMDB: Star Trek 2014) was initially broadcasted between 1966 and 1969. The actor William Shatner, who would provide the voice-over for the GRASS video in 1987, stars in the TV show as the Captain "James Tiberius Kirk" of the fictional starship USS Enterprise. The choice of William Shatner to narrate the promotional video was not accidental, as the actor Leonard Nimoy, who played the alien "Mr. Spock" from the fictional planet Vulcan in the same TV show, would have been the backup for Mr. Shatner. The launch of a sequel to the original Star Trek TV show in 1987, named "Star Trek: The Next Generation", without the original cast, seems to be coincidence (IMDB: Star Trek: The Next Generation 2014). From the early days of GRASS GIS development, user feedback to the developers had been influenced by the concepts of advanced information visualization as foreseen by the Star Trek TV show, thereby indirectly affecting the evolution of the GRASS software (Westervelt 2004). The striking similarities of the tolerance-based behavioral codes among the fan communities devoted to the Star Trek TV shows and the meritocratic values of OSGeo project communities remain to be analysed (Shatner & Kreski 1999, Löwe & Neteler 2014).

3 Reference Case: Preserving and Citation of the GRASS GIS Video

3.1 Non-textual Media at the German National Library of Science and Technology

The German National Library of Science and Technology (TIB) is one of the largest specialized libraries worldwide. TIB is a member of the Leibniz-Association, a German umbrella organisation for 86 institutions conducting research and providing scientific infrastructure. It is jointly financed by the federal government and the federal states of Germany. The TIB's task is to comprehensively acquire and archive literature from around the world pertaining to all areas of engineering as well as architecture, chemistry, information technology, mathematics and physics.

Within TIB, the Competence Centre for non-textual Materials is committed to improve the access and use of non-textual material ranging from audiovisual media, research data to 3D objects. This material is to be systematically collected and preserved as cultural heritage. For this task, an advanced web-based platform for audiovisual media (AV-Portal) is currently being developed by TIB and the Hasso-Plattner Institut for software system technology GmbH (HPI)(Neumann & Plank 2013).

The AV-Portal optimizes access to and the use of scientific videos from the fields of engineering and science in the face of the rapidly growing numbers of scientific film being published on Web 2.0 platforms.

For this, advanced multimedia analysis methods such as scene, speech, text and image recognition are combined in order to enhance the bibliographic metadata to enable extended search capabilities. The results are also connected to new knowledge by linking the data semantically. The aim is to make it as easy for users to locate and use the growing stock of non-textual material as it is for them now to procure textual media.

In addition, films stored in the AV-Portal are assigned with digital object identifiers (DOI) as persistent identifiers to ensure that the non-textual media are accessible long term from difference sources, irrespective of their current location, enabling long term citation and referencing. This enables fine-grained citation using the Media Fragment identifier (MFID) standard to provide a individual citable DOI for each segment of a film.

The AV-Portal will be released by mid 2014, providing acces to scientific films in German and English.

3.2 Discovery of an Alternative Version of the GRASS Promotional Video

Because the problems regarding the citation of the GRASS promotional video were known during the testing phase of the AV-Portal, the original copyright owners of the video were contacted, whether it could become a test case for the AV-Portal media collection. Following an initial positive response, the following steps triggered a search in the archives of the original producing company, Moving Pictures Productions. In this process, a formerly unpublished high resolution version of the GRASS promotional video was discovered. The content of this high resolution video is currently being transcoded by TIB and will undergo subsequent advanced multimedia analysis to receive enhanced, comparatively fine granular metadata which will be also indexed next to the bibliographic metadata. The searchable and citable video will become available online following the release of the TIB AV-Portal.

3.3 Upcoming Research Activities

Visual comparison of the previously known VHS-copy of the GRASS promotionaoe video and newly discovered version show that multiple differences exist. An example is given in Figure 2. At this point it is assumed that the high resolution footage is an earlier edit and predates the VHS-version. An in-depth MFID-based comparative analysis will be conducted once both videos have been published on the AV-Portal.

Figure 1: Example of the improved coloring and video resolution of the newly discovered high resolution version of the GRASS promotional video (right) compared to the VHS-version (left).

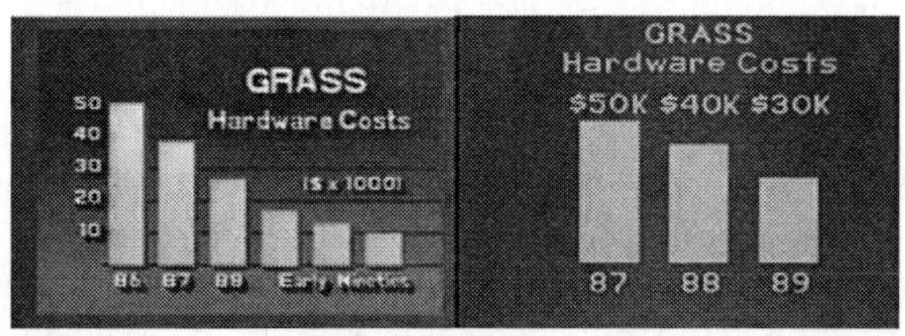

Figure 2: Alternative content in the previously known version of the GRASS promotional video (left) and the newly discovered high-resolution version (right) .

4 The Road Ahead

The preservation of AV-content as described for the GRASS promotional video is a crucial first step, but it is only a part of a larger unified preservation effort for research data on a larger scale: OSGeo Projects are centered on the development of software code, which can in turn become a field of analysis and research, taking advantage of the twin nature of software being perceived as data and vice versa (Graham 1995). For this, visualizations and animations are already being used and published on Web 2.0 portals, which contain the described limitations regarding citability and long term preservation. Figure 3 provides an example on the current state of the art. However, the tight coupling of multimedia, research data and software repositories enabling reference and citation via DOI remains an active research topic for non-textual information management.

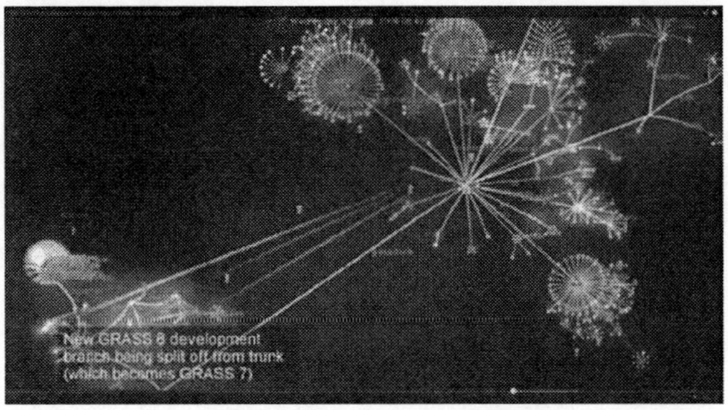

Figure 3: Still frame from a visual analytics animation providing a high level view on the evolving of the GRASS GIS codebase using the Gource software for software version control visualization (Neteler 2013, Grouse 2014). .

5 Conclusion

The integration of the historic GRASS GIS promotional video from 1987 into the upcoming TIB AV-portal is a success story and will serve as reference case for future multimedia preservation activities within the OSGeo communities. The discovery of an unpublished alternative high resolution version of the footage will enable further research using DOI-based citation of the video versions.

While this is a rare and unexpected find from the early days of Geoinformatics, the benefit provided by web portals such as the TIB AV-Portal to the OSGeo project communities lies in preservation, improved access and citability of contemporary multimedia information: The supply, use and significance of non-textual media is continually increasing while only a tiny proportion of these materials can be searched and explored right now.

One response to face these new challenges is the extension of library portals to accommodate non-textual information, develop new tools for indexing, searching, browsing and displaying the data, including software, as well as enrich the data with semantic information. The new search services for library user communities, including the OSGeo project communities, will provide innovative search scenarios and new ways of tapping into knowledge.

References

Graham, P. (1995) ANSI Common Lisp Book, Prentice Hall

GRASS GIS (Last checked 2014), 'The world's leading Free GIS software'.
URL: http://grass.osgeo.org/

GROUCE Software (Last checked 2014),
URL: https://code.google.com/p/gource/

IMDB: Star Trek (Last checked 2014). URL: http://www.imdb.com/title/tt0060028/

IMDB: Star Trek: The Next Generation (Last checked 2014). URL: http://www.imdb.com/title/tt0092455/

Löwe, P., Klump, J., Thaler, J. 'The FOSS GIS Workbench on the GFZ Load Sharing Facility compute cluster' (Last visited 2014), URL: http://presentations.copernicus.org/EGU2012-4491_presentation.pdf

Löwe, P., Neteler, M. (2014), 'Data Science: History repeated? – The heritage of the Free and Open Source GIS community', Geophysical Research Abstracts, Vol. 16, EGU2014-10027, 2014

Mason, B (2013). 'Video Flashback 1987: Shatner Tells You Where You Can Stick Your Maps' (Last checked 2014). URL: http://archive.wired.com/wiredscience/2013/08/shatner-loves-digital-maps/

Neumann, J., Plank, P. (2013) 'TIBńs Portal for audiovisual media: New ways of indexing and retrieval' URL: http://library.ifla.org/92/1/124-neumann-en.pdf

Neteler, M. 'GRASS GIS 6.4 development visualization from 1999 to 2013 (with soundtrack)' (last checked 2014). URL: https://www.youtube.com/watch?v=MR4_5GSID2A

Neteler, M. 'Scaling up globally: 30 years of FOSS4G development' (Last checked 2014). URL: http://de.slideshare.net/markusN/scaling-up-globally-30-years-of-foss4g-development-keynote-at-foss4gcee-2013-romania

OSGeo (Last checked 2014),'Your Open Source Compass'. URL:https://www.osgeo.org/

OSGeo: GRASS MOVIE CERL (Last checked 2014) URL:http://grass.osgeo.org/grass_movie_CERL_1987

Shatner, W. & Kreski. K. (1999), Get A Life !, Pocket Books, New York, USA 1999

Spearfish Sample Data (Last checked 2014).
http://grass.osgeo.org/sampledata/spearfish_docs_1979_p163to171.tar.gz

Westervelt, J. (1991), 'Why does GRASS Exist (Excerpt from "Introduction to GRASS4")' (Last checked 2014) URL: http://grass.osgeo.org/grass41/grass1to4history.html

Westervelt, J. (2004), GRASS Roots, in 'Proceedings of the FOSS/GRASS Users Conference (Bangkok, Thailand, 12-14 September 2004)'.

Youtube: GRASS MOVIE CERL (Last checked 2014) URL:https://www.youtube.com/watch?v=U3Hf0qI4JLc